"The web never quite lived up to its lofty promises, and now it's entered a state of widespread decay. But to everyone who came of age, found work, or met their people online, there's still much to savor and much worth saving. In this haunting, personal, and singular book, John West explores the ways that the digital sphere collects and contorts our personal histories and binds them up in the fates of programs and platforms. After all, those of us who have grown up with the internet must ultimately grapple with the fact that we will die with it."

—**BRIAN MERCHANT**, author of *Blood in the Machine: The Origins of the Rebellion Against Big Tech*

"Most writing about the dying internet is either nostalgic or doom-y. John West has found a third register: clear-eyed, alive, even quite funny. A book about loss that's somehow a pleasure."

—**PJ VOGT**, host of *Search Engine*

"When you're a kid, you think what's on the internet will last forever. When you're an adult, you realize that's a lie. This book is a lovely meditation on what it means to reckon with one's own mortality, the impermanence of the digital, and the mysteries of faith."

—**MEREDITH BROUSSARD**, NYU professor and author of *Artificial Unintelligence: How Computers Misunderstand the World*

"The internet only ever captures more and more of our time and attention, yet everyone struggles with the way it leaves us feeling empty. Calling on familiar traditions and profoundly new insights, John West offers us a way to find the meaning and soul that's missing in today's technology."

—**ANIL DASH**, technologist and writer

"At once informed and panoptical, John West's *The Internet Will Die, and So Will You* probes the mutability and ephemerality of the digital systems that now enfold us. As the title suggests, he widens his reckoning to include us, their mutable and ephemeral users."

—**SVEN BIRKERTS**, author of *Changing the Subject: Art and Attention in the Internet Age*

"Remember when wasting time on the internet felt like a choice you were making—with all the joy and guilt of a bad decision consciously taken? Remember when you had to remind yourself to check your email? John West's lovely book won't give you back all the hours that you let the infinite scroll steal from you, but it will restore a little of that consciousness of choice. Like the best Web 1.0 sites, this book is strange, intelligent, and gratifyingly hard to classify."

—**PHIL CHRISTMAN**, author of *Why Christians Should Be Leftists*

THE INTERNET WILL DIE, AND SO WILL YOU

ON LEARNING TO LIVE IN OUR DIGITAL AGE

JOHN WEST

HERE BELOW BOOKS
GRAND RAPIDS, MICHIGAN

Here Below Books is an imprint of
Wm. B. Eerdmans Publishing Co.
2006 44th Street SE, Grand Rapids, MI 49508
herebelowbooks.com

Published 2026

Printed in the United States of America

32 31 30 29 28 27 26 1 2 3 4 5 6 7

ISBN 978-0-8028-8542-5

Library of Congress Cataloging-in-Publication Data

A catalog record for this book is available from the Library of Congress.

Quotations of Scripture marked KJV are from the King James Version. Those marked NIV are from the New International Version.

CONTENTS

Our days are like the grass.
We flourish like a flower in a field.
Until the wind blows, and the flower is gone.
Not even the soil will remember it.

—Adapted from Psalm 103:15–16

Part 1

THE INTERNET WILL DIE . . .

A word is dead
When it is said,
Some say.
I say it just
Begins to live
That day.

—Emily Dickinson

CHAPTER 1
INTERNET

Fire in the hole

On Wednesday, July 18, 2001, the heat that had gripped Baltimore for days finally broke. The downtown office workers sneaking out for their midafternoon smokes were, at last, unbothered by sweat and sweltering humidity. Beneath their feet, unconsidered, the Howard Street Tunnel stretched for nearly two miles. Hailed as "a railroad triumph" by the *Baltimore Sun* when it was built in the 1890s, it took 2,400 men five years to finish it. Now able to accommodate double-stacked trains, the tunnel still transits vast quantities of freight to and from the northeastern United States. But more than trains run through its brick-lined walls: Several internet service providers and telecommunications companies run cabling through it.

A little after 3:00 p.m. that Wednesday, a sixty-car train derailed in the tunnel. Flammable liquid ignited, causing a conflagration, according to a subsequent National Transportation Safety Board report. The temperature in the tunnel reached 1,000 to 1,500 degrees. "We're told some of the boxcars are actually glowing," a fire department spokesperson told the Associated Press. Over one hundred firefighters would be called up to fight the blaze. No one was killed, though five emergency responders suffered minor injuries. The NTSB never did figure out what went wrong.

The freight company, CSX, rerouted other trains to avoid the inferno, but the internet couldn't be so easily retransmitted. The cables that wound their way through the tunnel were damaged, and web traffic across parts of the country slowed to a crawl. "Keynote Systems, which measures the performance of Web sites, said slowdowns in Atlanta, Seattle, Los Angeles and elsewhere were probably related to the burned cables," the Associated Press reported. "The company said the internet delays were the worst it had seen." A single tunnel—a choke point—captured much of the country's web traffic, and the country as a whole paid the price.

The internet isn't what we think. It is physical, subject to the vagaries of nature and time, just as our bodies are. Like us, it will flicker and fade, defined as it is by its fragility, which crops up everywhere: in the sudden disappearance of a relied-upon website, in the ever-present thrum of posts, images, and videos that accumulate seemingly without end. This is a book about that fragility—and what we do with it.

In the case of the Howard Street Tunnel, the fragility became, without warning, quite obvious. In their essay "We Need to Rewild the Internet," published in *Noema*, Maria Farrell and Robin Berjon highlight the role of WorldCom—a then-major phone and internet company—in the outages that followed the inferno. "WorldCom's resilience planning meant traffic was spread over different fiber networks in anticipation of just this kind of event," they write, citing MIT professor David Clark. But "the region's narrow physical topography had concentrated all those different fiber networks into a single chokepoint, the Howard Street Tunnel. WorldCom's resilience was, quite literally, incinerated. It had technological redundancy, but not diversity." Despite dealing in the physical world, in the fiber-optic cables that comprise the real, material internet, WorldCom had somehow forgotten that the land makes its own demands—that bridges and tunnels remain the only way to traverse the impinging waters of the East Coast. As Farrell and Berjon write, "Bottlenecks aren't always obvious."

Bottlenecks have their benefits, of course. There are efficiencies to be found in transmitting data through a single point, through companies that consolidate the services we need to house our lives online. These behemoths make a virtue of their size and importance. "Centralized service providers say they're better resourced and skilled at attacks and failures," Farrell and Berjon write. And there's money to be made in identifying these sorts of quasi-monopolies. "Widespread outages due to centralized chokepoints have become so common that investors even use them to identify opportunities," Farrell and Berjon continue.

In 2021, Fastly, a cloud utility provider, went down for an hour, taking major news outlets' websites—including CNN, the *New York Times*, and the *Guardian*—with it. *The Verge*, a technology news site, ended up publishing articles via Google Docs while its website was down. During that outage, rather than taking a beating in the market, the company's stock reached new highs. "Investors were delighted by headlines that informed them of an obscure technical service provider with an apparent lock on an essential service," Farrell and Berjon write. "To investors, this critical infrastructure failure doesn't look like fragility but like a chance to profit." Fragility and profit: They go hand in hand, which makes some sense when you remember that redundancy always carries inefficiency. Brittleness is big business, a feature and not a bug of the cloud economy. But you wouldn't know that if you take that word, *the cloud*—the metaphor of it—as true.

Partly cloudy with a chance of mixed metaphors

I suppose the metaphor *is* a good one, at least inasmuch as it clouds our awareness of the very real collection of companies, people, and processes that create and structure the internet. "Cloud storage requires paying someone, an outside entity, for as long as you are engaged in the act of storing," Maxwell Neely-Cohen writes for the Harvard Law School's Library Innovation Lab in an essay about the problems of "century-scale storage." When we put something in the cloud, we don't just offload the responsibility of maintaining it to fickle chunks of magnetic metal alight with electricity; we also offload that responsibility to a rather limited set of fickle human institutions. "Most companies do not last long," Neely-Cohen continues. "They get acquired or split up into pieces or go bankrupt or decline into something much smaller or are upended by catastrophic geopolitical events." An act of trust rests at the heart of cloud storage—trust in the same profit motive that thrives on fragility.

And that same metaphor—the cloud—also shrouds the cables, data centers, and transmission towers that comprise the internet. "The cloud is made possible by a relatively small number of undersea cables that require constant maintenance," Neely-Cohen writes. For a medium so often praised as decentralized, for a metaphor that calls to mind something diffuse, this is all an awful lot of centralization.

We miss the droplets of data in the mists of the cloud. "Digital copies of books, film, music, games, and journalism, some released under subscription models, some available on digital marketplaces, are all part of our cloud environment," Neely-Cohen writes. "Over the past decade, the collection development budgets of most libraries have moved steadily away from physical books and towards cloud-based digital subscriptions." In other words, a small number of companies own the data centers connected by a small number of cables, also owned by a small number of companies, through which data, yet again owned by a small number of companies, flows. The internet, in this telling, looks much more like the Howard Street Tunnel than a cloud. Indeed, the most cloud-like part of this state of affairs is the absence of human agency.

That's *our* human agency—we, the users of the web. Someone's attention is paid to the data and infrastructure the cloud cloaks. Hordes of flesh-and-blood engineers sitting in glass-and-steel office buildings and water-and-air-cooled data centers attend to the systems on those stacks of silicon that hold our words, pictures, and videos. What we store, what we transmit—these are all acts of attention to a world that remains resolutely material, despite the seeming immateriality of our digital lives. Human institutions remain central to the practice of storage and transmission, an inconvenient fact since human institutions fade, just as human bodies do. The cloud is refreshed every moment of every day with the fresh attention of humans. Just as manure, in balance, can encourage the dogwood to flower, so too does our care—over generations—make the internet bloom.

I mean this in a very literal sense, down to the ways the internet structures information. Take the book as example, which has mostly been recognizable as a way of structuring information for centuries, and compare it to the formats we use online. GIFs and HTML, chunks of PHP code and Java applets—these all require deep wells of technical expertise to hold the things we ask them to hold. But, generally, to pay attention to one computer language means we don't pay attention to another, just as to encode data one way means we don't encode it in another. The act of displaying a website on a computer or an app on a phone takes several layers of computers all talking to each other, all passing data formatted in particular ways. As the internet pushes innovation onward, new formats are being created all the time. The old ones must wilt without the attention of software engineers beaming energy into their leaves.

Flash in the pan

It's been several iterations of the internet since I learned to code in ActionScript, the language used by Adobe in its proprietary multimedia software platform Flash, which flourished in the early aughts and was used to make interactive web graphics and games. I learned the language to make myself more appealing to potential employers at the intersection of technology and news, since Adobe Flash made the early news sites' forays into data and web-native storytelling possible. Flash powered "those great, complex interactive graphics and data visualizations you see at the *Times* and elsewhere," Joshua Benton wrote at *Nieman Lab* in 2011. But Steve Jobs insisted that Flash wouldn't be supported on the early iPhone, and by 2011, Adobe ceased supporting new mobile Flash development. The writing was on the wall. Before long, the language was as dead as Latin or Sanskrit: spoken, perhaps, by a few specialists in special moments, but no longer a vernacular.

"Journalism is often considered the first draft of history, but what happens when that draft is written on a software program that becomes obsolete?" Clare Duffy and Kerry Flynn ask for CNN in 2021. What happens is news coverage of the September 11, 2001, attacks can no longer be accessed. "What was once an interactive explainer of how the planes hit the World Trade Center or a visually-rich story on where some survivors of the attacks are now [became], at best, a non-functioning still image, or at worst, a gray box informing readers that 'Adobe Flash player is no longer supported,'" they continue. They highlight *USA Today*'s 2002 investigation into the World Trade Center's elevator system "that included a Flash graphic explaining how people got trapped inside them." In order to keep the graphic alive, they write, "the USA Today team chose to remake that graphic and republished it." Note that prefix, *re-*. It isn't just the mediums that carry the data that must be ever refreshed; the way the data itself is encoded must also be refreshed with the sweat and tears of software engineers.

I am one of those software engineers. I am also a journalist, working for the better part of a decade to report the news with code at the *Wall Street Journal*. Over the course of my professional life, I've seen countless websites lose themselves in the cloud, and I've watched as others persist long past their expiration date. At the heart of these mismatches beats a misunderstanding of the web's physicalness and concreteness, and, thus, its relationship to time.

Having the time of our lives

We know, offline, that time demands choice. How will we spend it? One hobby to pursue and another to drop, one artifact to keep and another to toss, one moment to imbue with meaning, again and again, and another to pass by. To begin to bake means, necessarily, that there is less time to garden. Sticking your friends' wedding invitation to the refrigerator door covers your nephew's school photo. The day of your first date with your partner might deserve flowers, but you forget the day you moved into your first shared apartment.

Meanwhile, our web providers and platforms, like Odysseus's sirens, sing the honeyed lie: *You don't need to choose; we'll choose for you.* Create an account on Goodreads, and it remembers every book you feed it. Add a photo to your Instagram grid, and the grid grows larger. Each birthday for each "friend" on Facebook, no matter how distant, will be sent to your devices.

The desire to avoid choice, to offload it, is bound up with our annoyingly mortal bodies. Choice embodies the singular truth of the human condition: We are finite. Time is scarce; attention is precious; our lives are, in the end, frail and fleeting. I know I ought to spend this little life of mine in full possession of whatever agency I can marshal, but not choosing means I can ignore, if just for a little while, that I get a painfully few thousand weeks to make my choices—and that's if I'm lucky. So the platforms' lie, understandably, tempts even the best of us. The serpent said it best, long before Facebook did: "Ye shall not surely die."

Every time I think I've made peace with the limits of my own human life, something forces me to grapple, yet again, with that fuse that was lit at birth, whose sparks animate my aging body, which parades me down the wire to my inexorable death. I wouldn't trade the time I spend with my daughter for anything. Still, as my knees click each time I drop to the floor to play with her, I am reminded that my body grows older, that the time I thought was mine—hoarded in days spent reading or writing or watching television or TikTok—isn't mine at all. "Time is the fire in which we burn," as Delmore Schwartz writes in the poem "Calmly We Walk through This April's Day." Or, if you prefer a different metaphor, in Robert Hass's poem "Time and Materials," his speaker desires "to render time and stand outside / The horizontal rush of it, for a moment / To have the sensation of standing outside / The greenish rush of it."

Now, nearly forty years old, I can see with almost painful clarity the way time held back its sea-foamed, brackish rush in my teens and twenties. Naive, I thought that life would continue to stretch itself out for me. But life interposes itself between my time and my ambition. Work takes. The joys (and pains) of family demand. The ocean-view flats on which I imagined building my life, it turns out, were only borrowed—and death has begun its great, lifelong reclamation.

In my fitful morning meditations, as I muddle through my attempts at spirituality, I struggle to find peace with that fact. My beard changes color, my hair recedes, and my body softens in this autumn of my life, as I harvest the fruits planted in my twenties and thirties. My daughter's childhood, which I photograph with ever more desperation, rushes to its close. My parents continue to slow, become unmoored from my imagination of them. That fickle resource—human attention—structures everything as it flickers and fades: my rough prayers, the memories of my daughter's chubby face, the reality of my parents' aging. Impermanence stings. The grave will have its triumph.

The same all-too-human dispossession runs through the web. When we lose agency over our digital possessions, we lose a little bit of what makes us human—namely, the ability to memorialize and make meaning on our own terms. This is no different from when we lose our house in a fire, or when we're gripped by the compulsion to horde. In both these cases, the tragedy comes from the loss of agency as much as the loss (or accumulation) of the artifacts themselves. Despite the terabytes upon terabytes of storage, despite the lies of our platforms, our digital lives are finite because life itself is finite. I mean that in both an existential sense and a practical one. The users of the internet, we frail humans, will eventually die, but also, impermanence is baked into the very structure of the web. The sites we love and loathe will not last forever, despite how it may feel. The web is always changing, being remade before us, reliant as it is on human attention to structure it.

But finitude imbues our lives with grace and purpose, I have to believe—even online. My partner told me recently that if her best friend died, she would print out their text messages to each other, bind them into a book, and read them like literature. I found myself moved by this, by the idea that even the casual SMS finds thick meaning in death. Someday, I am going to die. It's unlikely to be soon, God willing. Maybe my partner, if she outlives me, will print out our text messages, full of tiny tasks and quotidian questions. Maybe those moments, too, will find meaning in my passing.

For a while now, I've snapped pictures of ephemera on my phone, which allows me to destroy their physical forms guilt-free. With the magic of artificial intelligence, I can search "grocery list" in my photos and find, immediately, scores of scrawled *apples*, *bananas*, and *chicken breasts* from years ago. These grocery list doppelgängers have continued to exist in a strange and purposeless half-life, shadows cast by their long-decomposed bodies. Unlike their pulp-and-ink versions, I feel no pressure to choose to bin them. At least not until I want to find a photo in my albums that actually matters.

It would be easy, these days, to live a life untroubled by the need to store our physical memories and artifacts. That verb, *to store*, comes from the Latin *instaurare*, which means something like "to set up and establish" or "to renew and restore." It is an active word, a continual action.

My partner and I have spent untold hours organizing and reorganizing our library of books—by color, by genre, by author's last name, and, most recently, by author's birth date—setting up and renewing our relationship with our most numerous type of possession. In this way, we've taken a leaf from Walter Benjamin's 1931 essay "Unpacking My Library," where the philosopher describes unpacking box after box of books, reliving important memories of his life, outlining a relationship to his books that "does not emphasize their functional, utilitarian value—that is, their usefulness." Instead, he thrills at "everything remembered and thought," which comes together to produce a "magic encyclopedia," one that he can—and does—hold pieces of in his hands. So, properly, *to store* does not mean *to forget*.

Online, we forget what we post all the time, though. I can't help but contrast Benjamin's magical and well-used encyclopedia with its notches and tabs to our digital lives, which make it so easy, so convenient, to hand off our artifacts to the mysterious and seemingly all-powerful cloud—to the abstract notion of an internet somewhere out there, all around us, where our words and photos, friends and family float deincarnated.

In their rhetoric and business models, in their interfaces and patterns, our platforms nudge us to digitize and offload our thoughts and experiences. If—or when—we're seduced, they decide what is stored and, thus, what is reproduced. But we only remember what we reproduce, what we perform again and again. In that performance, we are rebecoming what we were—etching deeper grooves into the records of our lives. We are what we remember because we remember what we attend to, what we perform. This means that when we lose the choice to neglect or to attend of our own free will, we lose agency over who we can be. The story of the internet in its modern incarnation is exactly that: a shift from human agency to platform agency.

A tale of two websites

Presaging the nerd I would flower into, I built my first-ever website in middle school. It was a fan page dedicated to Japanese animation—chiefly the film *Princess Mononoke* and the television show *Cardcaptor Sakura*. That was over twenty years ago, and though I've since tried to take it down, it still exists, gaudily pink and green. The ancient AOL email address associated with the account—a misspelling of a mostly forgotten German baroque composer—has long since been deleted, but its code remains on some server very far away, on some blinking machine in a windowless warehouse full of other, indistinguishable blinking machines, which themselves carry blueprints for other websites about other terrible television shows.

But there's that word again, the one we use for the detritus generated by our lives, for artifacts meant to exist only for the moment of their use: *ephemera*. A paper receipt from the corner deli, a grocery list scrawled on the back of a receipt, my old website. I should say that generally I enjoy ephemera, especially the dog-eared, battered kind. Each wadded-up movie ticket and drink-stained concert bulletin opens a window into a past version of myself. It often feels wrong, somehow, to dispose of these little reminders of myself, the old ways I used to be, the old things I used to love. Binning them would be like disposing the reminders of living—disrespecting the beautiful and mundane activities of just being.

But my website: It might seem odd to call a website ephemera, but I didn't mean mine to be a monument. I built it mostly to signal to my friends that I was serious about anime and to try to make new friends in that first wild iteration of the mass-market internet—though on neither count did my little page succeed. Like most ephemeral objects, meant to be disposed of, meant to be useful rather than beautiful, my website was forgettable, and by the time high school rolled around, I had mostly forgotten it.

Neglect may be the ultimate enemy of our mementos, but despite my neglect, that website still works—for the most part, at least. Blinking advertisements for seedy services take up more and more of its real estate. The now-archaic way I coded the site, using a feature called "framesets," isn't fully supported by modern browsers, so its images and text misalign. A few pages I know I once labored over are mysteriously unreachable—probably deleted—and the "guestbook," where friends or strangers would write a few words, is also gone. Despite these minor failings, against all odds, my silly adolescent website persists. You can still see my terrible fan art and read my tiny, incoherent essays about environmentalism. They still spring to life every time its address is entered or the exact search result is clicked.

This website's continued existence haunts me. Not because it's particularly embarrassing; I'm fond enough of the kid who planted his tiny flag in the *terra nova* of a digital world. No, it haunts me because it lives on despite my neglect. I should be able to be rid of it. Instead, the words I wrote in middle school still sound in the minds of the random readers who stumble upon them. It's like a witch cursed me, and now I have to recite, "The website is deticated [*sic*] to a very spisific [*sic*] genre of anime/manga: EnviroAnime" to anyone who knocks at the right door.

I created my second website—a blog this time—a few years later, in college. *Ich Bin Ein Oberliner* never welcomed more than a few hundred monthly visitors, but I wrote with a great seriousness that somewhat belied the punny name—a play on my college's name and a famous malaprop by President Kennedy. (I must credit the wordplay to my friend Lindsay.) I put out a few posts a week (spell-checked, thank you very much) about whatever interested me. All around me, the web was blossoming, it seemed. Like the smell of lilacs in spring, ideas were in the air, and I could add to the blooming garden. I still remember the exhilaration of this: writing in response to another, and reading what they wrote in response to me. My blog lived on Google's blogging platform, but in 2011, in a fit of DIY optimism, I moved my blog over to WordPress, a popular, open-source piece of blogging software, which I customized with my own code and design. The catch was that I needed to host it myself—that is, pay for server space and a URL. At some point I stopped paying the bills. That last year of my blog disappeared—neglect is the ultimate enemy of our mementos—and the domain now points to a German casino, if Google Translate can be trusted. The early years still exist on Google's Blogspot, though they are a shell of their former glory. Error messages replace images, and links point to purveyors of spam, scams, and malware.

Unlike my anime page, I never wanted this blog to disappear, though I have no excuse for my neglect. I miss it. Not the words, exactly. I miss that year. I miss the *me* it represented. The early blogosphere was fertile ground, a fecund place where writers, including me, grew into themselves, all thick bark and leafy canopies, which brushed against their neighbors—idyllic. Those were halcyon days. I mean that in the common way people mean it—that they were glorious, though tinged with bittersweet nostalgia—and I mean it in a deeper way too. That phrase, *halcyon days*, comes to English from a myth, referring to the uncanny calm a wind god created so his kingfisher daughter—a halcyon bird—could lay her eggs with her mate. Before she was a bird, she was a human. She threw herself into the sea when her lover was killed in a storm, but the gods saw fit to transform them both. In other words, the internet's halcyon days were never more than a brief interruption of modernity's ongoing storm. I know this. I still miss who I was when I wrote my blog: my college self, young and hopeful. It was, after all, the best of times—unlike my middle-school days when I created my *Cardcaptor Sakura* site, which were basically the worst.

So, one website withers when I want it to flourish. The other sends up shoots when all I want is to pull it from the soil. A story of the internet in two parts: First, the things you wish would disappear last forever, and second, the things you think will be reliable flicker then fade.

We remember best when we reenact the past. We used to remember Homer because we repeated his story, word for word, hexameter for hexameter, reproducing it in our throats. Now we remember it because, every few years, someone takes it upon themselves to retranslate it, reproducing it in our printers, bookshelves, and minds. We still remember Bach's cantatas because we have spun up entire institutions, conservatories and orchestras, around their reperformance. Every year, my Jewish friends eat matzo, bitter herbs, and salt water to feel the Passover story burning in their throats. The nostalgic among us still remember the websites they loved because they revisit them, demand their browsers re-create them again and again.

My kingdom for a click

My blog posts nearly always responded to others. They were just one side of a conversation. Another's post, another's idea, another's art, another's reference—these formed the backbone of my blog's subjects. In other words, blogs reproduce themselves socially. Even if I could somehow undelete that last year, I still couldn't reenact it—not really. The conditions in which it thrived have ceased to exist—have been magnetically overwritten as thoroughly as the hard drives that used to house *Ich Bin Ein Oberliner*. A vibrant culture of conversation moved to other parts of our digital world. In part, blogging professionalized and merged with legacy media; once the *Washington Post* and the *New York Times* started blogs, the end of independent blogging was all but certain. After all, there's money and prestige to be had at the *Post* or the *Times*—features an independent blog (even the word *blog*, so diminutive!) decidedly lacks. Consider Ezra Klein, an early blogging pioneer. He wrote his way to a job in the aughts blogging at the *American Prospect*, a respected, if small, left-of-center magazine, then again at the *Washington Post*. He left the *Post* to create his own digital publication, *Vox*, and finally, he landed at the *Times* as a podcast host and opinion writer. Not every blogger followed that trajectory, from outside the walls of professional media to smack-dab in the middle of it. The fact that it was possible for a few made it tempting for many. In that way, the great skimming of the cream of the blogosphere, which meant success for the blogosphere's best and brightest, also weakened the old incentive structure of nonprofessional, independent bloggers.

The author Nicholas Carr, on his now-inactive blog *Rough Type*, saw the coming collapse even in the early days of blogging. In 2008, he wrote that "almost all of the popular blogs today are commercial ventures with teams of writers, aggressive ad-sales operations, bloated sites, and strategies of self-linking. . . . To argue that they're part of a 'blogosphere' that is distinguishable from the 'mainstream media' seems more and more like an act of nostalgia, if not self-delusion." Carr points to a "state of the blogosphere" report released by Technorati—which used to be a search engine for blogs and which has long since pivoted to simply selling ads—that is no longer easily accessible on the web.

Other factors contributed to the blogosphere's decline. The smartphone became the dominant way people access the internet sometime in the teens, according to research from the analytics company StatCounter. And the way people access the internet on their phones, by and large, is through apps, not browsers—meaning readers are less likely to enter the URL of a blog's home page than to scroll through Facebook or X or TikTok. You might see a link to a blog post in your Facebook feed. You might even click on it and make your way to the post in question. But this "side-door traffic," as it's called, replaced loyal readers with fickle ones. While over 80 percent of readers remembered where they read something when they found it directly, less than half could correctly remember where they read it when they came from social media, according to a study from the Reuters Institute for the Study of Journalism at the University of Oxford. A necessary precondition for loyalty, it seems to me, is remembering where you read the things you read. Facebook and its kin mediate their users' relationships with the links they spread by diminishing what makes those links unique. Each post has the same weight, the same form of a big image with a headline over it and the publication listed in little gray text at the bottom. This is what drives us to say things like, "I saw that on Facebook," rather than "I saw that on *Ich Bin Ein Oberliner*."

The move to a mobile-first internet coincided with the rise of "walled gardens"—platforms that aimed to keep users on their site and only their site. The two are related. It's hard to multitask on a phone, so there's a good chance that if you let a user leave for a moment, they'll be gone forever, while on a laptop, you can click and open a link in a new window. These walled gardens also needed to make money. Independent bloggers might have been supported through subscriptions, donations, or ads—though as often as not, they were simply writing as a labor of love. But platforms need revenue that scales alongside their user base. This has meant, generally, ads. And to more easily shove ads down our eye gullets, platforms need us to stay on their site—not follow a bunch of links to someone else's content.

Sitting in my off-campus studio apartment at the height of my blogging days, every morning, I fired up my blog reader: a website that aggregated all the blogs I read, putting them all in one place. The technology that enabled this aggregation is called RSS, an acronym that can stand for Really Simple Syndication, Rich Site Summary, or RDF Site Summary—RDF being yet another acronym. Software engineers aren't known for their Really Sophisticated Sense of humor, famously.

Today, to decide what we read and in what order, our social media platforms use algorithms powered by sophisticated statistics—called "machine learning," or sometimes, "artificial intelligence"—but RSS helped us use a much older kind of algorithm to decide what we read and in what order: human discernment. Social media platforms have incentives to nudge us toward certain kinds of content; they have opinions about what we should read, which they enforce in their design and code. Software backed by RSS is agnostic about how we chose to sort and sift; they ask humans to tag posts, to bookmark them, to read them on their own time and in their own way. An algorithm, after all, is just a fancy word for a set of instructions for accomplishing some task, and human algorithms—"This title seems interesting" or "I've liked posts from this author before"—have served us well in the past, and they still serve us well in some places online.

While RSS has fallen out of fashion (or been pushed out the window) for sharing text- and image-based posts, it still exists today under the hood of your podcast app. RSS, the technologist Anil Dash noted in a post on his (still-active) blog, remains the radical force behind the phrase spoken at the end of so many podcast episodes: "You can subscribe to this show *wherever you get your podcasts*." Dash writes that this phrase represents "the triumph of exactly the kind of technology that's supposed to be impossible: open, empowering tech that's not owned by any one company, that can't be controlled by any one company, and that allows people to have ownership over their work and their relationship with their audience."

That word, *empowering*, hits the nail on the head. Human algorithms, which technology enables, which allow a person to choose what and when they want to read, empower us. Our platforms' algorithms, which prey upon human attention, which rely upon computers doing sophisticated statistics to predict what will rile you up enough to keep you glued to your screen, empower the platforms.

Those mornings in my apartment, I would enter the blogosphere through Google's free RSS reader, which was named, without much imagination, Google Reader. Google shut it down in 2013, blaming declining use—a symptom and a cause of the decline of the open blogosphere. That declining use might be part of the story, but David Pierce reports at *The Verge* that Google Reader was never really resourced to succeed. By 2011, Google cut Reader's engineering team to the bone and left it on life support—in "maintenance mode"—meaning there wouldn't be any new updates, only bug fixes. Google was in the midst of trying to pivot to its ill-fated social network, Google Plus, and the company didn't want—had maybe never really wanted—to support Reader. After all, Google, too, wanted to build a walled garden, and it was laying the bricks of it right when it killed off its open Reader that embodied the spirit of the open web.

All these shifts left blogs in a strange position, culturally. As Addison Del Mastro writes at *The Bulwark*, "Blogs nowadays are a little bit like cathode-ray tube televisions—obsolete, still out there in large numbers." Social media shifted the cultural center of gravity online. Blogs still exist, of course, but they don't mean what they used to. "'Blogging' has always had two very different definitions," Carr writes. "One is technical: a simple system for managing and publishing content online, as offered through services such as WordPress, Movable Type, and Blogger. The other involves a distinctive style of writing." That writing style—and its place in the culture—has atrophied.

The disappearance of that last year of my blog is a convenient way to mark the end of my personal blogging era. My neglect, the blog's proximate cause of death, was itself caused by the trends toward professionalization and toward the mobile-friendly closed networks of social media platforms—by a lot of people and institutions with a vested interest in making me neglectful. So in crucial ways, the choice to neglect or attend wasn't really mine at all. It's easier to swim with the current than against it, after all. Easier to roll down a hill than trudge up it. Easier to use others' platforms than to maintain my own.

When websites rot away without our consent or when websites exist long past when we want to see them go, it's not just that we're losing some words and images and ideas, some mementos and ephemera, and that we're stuck with others. No, we're losing choice over how to be, over how to define ourselves. That is the essence of attention, those choices. As the web has shifted, our attention has shifted with it—to places where it cannot find the purchase that enables the kind of deep engagement necessary for human flourishing. We need a new way of thinking about attention online, one that takes into account our alienated stuff and our alienated selves.

But that new way might really be an ancient one. Every generation of humanity grapples with the fact of their inevitable deaths, with the necessary, poignant choices forced upon us by this all-too-human condition. Some fight against death. Some come to accept it. The same question—what is to be done about the fact of our mortality?—animates our digital selves. We use the web to practice dying, even when we don't realize it. Still, even there, the covenant from the beginning binds: "Thou shalt surely die."

CHAPTER 2
LIBRARY

Checking out

At its height, the Library of Alexandria—the Great Library—held tens of thousands of scrolls, a wonder of the world. It burned to the ground long ago, the story goes, taking with it much of the ancient world's knowledge, and leaving us with a too-tempting metaphor for the web.

Here it is, delivered adeptly by the journalist Adrienne LaFrance in the *Atlantic*: "The promise of the web is that Alexandria's library might be resurrected for the modern world." LaFrance was talking about the Internet Archive—which describes itself as a "digital library of Internet sites and other cultural artifacts in digital form." The archive aims "to provide Universal Access to All Knowledge." Its founder, Brewster Kahle, chose an old church as its headquarters in San Francisco, because "the church's front resembled the Internet Archive's logo: the Library of Alexandria's Greek columns," *Atlas Obscura* reports—a fact I confirmed with the Internet Archive leadership when I visited their building.

And the library metaphor continues to infect our understanding of the web. Harvard law professor Jonathan Zittrain notes in a June 2021 piece for the *Atlantic* that an informative page created by Google, titled "How Search organizes information," argues that "the web is like an ever-growing library with billions of books and no central filing system." The library metaphor might be less apt than it seems. As of December 2021, that Google page was titled "How *Google* Search organizes information" (emphasis mine), and it had dropped the reference to an ever-growing library. Then, in January 2022, the phrase made its return. By 2025, the page had begun bragging, "Google looks through hundreds of billions of webpages and other content stored in our Search index to find helpful information—more information than all of the libraries of the world."

Google, it seems, can't quite work out what its relationship is to libraries, metaphorically speaking. Is it a library? Is it better than a library? Is it a superhuman librarian fetching you exactly what you need in the exact moment you need it? But as even the descriptions of the library change and warp, I have to wonder, what kind of library would let the artifacts in its collection be so vandalized? LaFrance continues, "Today's great library is being destroyed even as it is being built." Online, even the metaphor of a library isn't stable. Or maybe the references to Alexandria's great, lost library are apt. Nothing lasts forever—not even online.

Rotten to the core

The vexing miracle of my long-lived first-ever fan site notwithstanding, most websites just don't live that long. The average lifespan of a website is only around one hundred days, according to Jill Lepore at the *New Yorker*, and all those dead sites have grave (pardon the pun) implications. One study found that half of all links contained in Supreme Court decisions are broken. Seventy-five percent of *Harvard Law Review* links are gone. Fifty percent of the *New York Times*' deep links—links to a subpage as opposed to links to a home page—have disappeared. Three out of four citations in scholarly articles have drifted—the websites they purport to cite no longer say what they once did. Over half of Wikipedia's entries contain at least one link in its references section that goes nowhere, a Pew report found.

This rot is caused by the very invention that makes the internet the internet: the hyperlink. Those blue, underlined bits of webpages that point to other webpages make the *web* in World Wide Web more than a metaphor. Take the Wikipedia page for "Human," which, as of 2023, opened, "Humans are the most common and widespread species of primate." Both *species* and *primate* are underlined and colored blue. Click on either, and you're taken to pages with entries for "Species" and "Primate," respectively. Click on any of the underlined blue links on either of those pages, and you're taken to yet more entries, and on and on. Those hyperlinks knit Wikipedia's many millions of English-language entries to each other in a complex, ever-growing tapestry—a smaller web within the broader hyperlinked network of pages that comprise our internet.

It's well known (Wikipedia even has a page for it) that clicking on the first hyperlink in any random article—and continuing to click on the first link for each subsequent page—will eventually take you to the page for "Philosophy." Enterprising readers have even invented a game in which players race to navigate from the page for one random concept—"Human," for example—to another, like the page for the color "Blue."

The hyperlink transforms text into hypertext—a series of documents connected in an immediately accessible way. Hypertext, in turn, lends its name to the basic code of the internet: HyperText Markup Language (HTML). Today, hyperlinks are no longer always styled as underlined blue words. They can look like images or buttons or whatever a designer's mind can dream up. Regardless of how they appear, the hyperlink, like a spider's dewy filaments, captures our insectile attention. But there's tension built into the very threads that make up the web: It's impermanent by design.

A website's owner can add, remove, and change the code stored at any time. There is no guarantee that the Wikipedia entry for "Blue" will exist tomorrow, nor is it certain that it will have the same language from day to day. If the page for "Blue" disappears, then those pages that link to it now link to an error page. If the page for "Blue" contains a mistake, then those pages now link to that misinformation. These additions, removals, and changes are silent: Websites don't alert their linked counterparts that their code has been updated. In other words, when we hyperlink one page to another, the linker is at the mercy of the trustworthiness and prosociality of the linkee.

Of course, people are untrustworthy—and, maybe especially in the age of social media, decidedly antisocial. So, more and more, as websites change and pages disappear, gaping holes form and grow in the web's very structure. Researchers call the outright disappearance "link rot" and the changes "context drift." It's frustrating, as a user, to find a site replaced with an error page. It's even worse when the lost citation is what powers a court decision or a scientific finding. It's even worse than that when instead of an error page, you discover something else altogether—something innocuous if you're lucky, and spam, malware, or misinformation if you're not.

Flexibility powers the web; flexibility also unwinds it. In a 2008 post titled "Internet Triumphalism," the pundit and early blogosphere wunderkind Matthew Yglesias noted that bloggers' ability to hyperlink directly to source material gave them an advantage over the old newspaperperson, who must rely on vague summarization. But, as links rot away, newspapers printed on dead trees will have the last laugh. Indeed, the example link Yglesias gives—to a report from a think tank—now binds his vintage take to the Brookings Institution error page.

Zittrain notes that during the Obama-era government shutdown, myriad "dot gov" sites broke, leaving curious taxpayers confused about what the government was doing. He also points out that, once, Justice Samuel Alito, in a Supreme Court decision, linked out to a page, which would later be made to read, "If you had [cited this webpage], like Justice Alito did, the original content would have long since disappeared and someone else might have come along and purchased the domain in order to make a comment about the transience of linked information in the internet age." Cute.

Jill Lepore notes that David Cameron, then a member of the British Parliament (he would go on to be the prime minister), argued that the internet was "making more information available to more people," and providing "the power for anyone to hold to account those who in the past might have had a monopoly of power." But, she writes, "seven years later, Britain's Conservative Party scrubbed from its Web site ten years' worth of Tory speeches, including that one." Later, Lepore continues, "BuzzFeed deleted more than four thousand of its staff writers' early posts, apparently because, as time passed, they looked stupider and stupider. Social media, public records, junk: in the end, everything goes."

Law and science, news and culture—these are all activities central to human civilization. Link rot threatens them all. It's like if Moses had come down the mountain not with a couple of stone tablets but with several billion Etch A Sketches.

The times, they are a-changing

But the danger isn't only that things we rely on might disappear. It's also that the world has the annoying habit of conforming to how we describe it. That is, as we describe our circumstances differently, those circumstances change: History, as the old saying goes, is written by the victor. So is the present. And, now, both can be rewritten by anyone with access to the servers on which history is stored—shaped by the hyperlink and the silently changed words.

It would be difficult, if not impossible, for a newspaper to alter the record contained in every one of the physical objects it printed in order to reflect some new reality. "DEWEY DEFEATS TRUMAN" can't be taken back—though I don't doubt the editors of the *Chicago Daily Tribune* wanted to try. Paper can burn, of course, but it's harder to edit. Online, though, a few keystrokes can alter the public record. And despite the taboo against rewriting the past, even storied journalistic institutions engage in so-called stealth editing—updating their headlines and stories without notice to the reader.

In 2016, Liz Spayd, the then-public editor of the *New York Times* (a position that the *Times* has since eliminated), reported that even the stalwart Gray Lady has taken to the practice. She noted that in an obituary for celebrated scientist Yvonne Brill, the *Times* cast her "in the first paragraph as a woman who 'made a mean beef stroganoff, followed her husband around from job to job and took eight years off to raise three children.' In secondary fashion, it mentioned that Brill was a pioneering scientist who invented a propulsion system that kept communications satellites in orbit." Spayd continues, "Seven hours later, after infuriated readers let loose, that version disappeared and a new story popped up with a less chauvinistic perspective emphasizing her exceptional career. No editor's note or other explanation for the changes was given." Future readers, it seems, would never see the *Times*' initial, retrograde foray.

Spayd notes that, as of 2016, she had written about two other cases of stealth editing, and her predecessor had written about seven. There were undoubtedly more instances where editors changed the news after its initial publication. I don't doubt that most "stealth edits" were neither cover-ups nor nefarious but rather small changes made to keep a news article up-to-date—a boon to readers, really. That is the argument that defenders of the practice make. First, readers ought to be able to see the most up-to-date version of the news. Second, unless it's a factual error or something extraordinary, readers just don't care. But what counts as up-to-date is increasingly up for grabs, and many readers, it turns out, do care.

Margaret Sullivan, Spayd's predecessor, recounts the saga of an article about Bernie Sanders's 2016 presidential candidacy, which initially carried the headline "Bernie Sanders Scored Victories for Years via Legislative Side Doors." The piece, Sullivan writes, "stayed in essentially that form for several hours online—with some very minor tweaks—but in the late afternoon, *Times* editors made significant changes to its tone and content, turning it from almost glowing to somewhat disparaging."

The editors changed the headline to "Via Legislative Side Doors, Bernie Sanders Won *Modest* Victories" (emphasis mine). That's somewhat less glowing. And the editors added some new copy as well: "But in his presidential campaign Mr. Sanders is trying to scale up those kinds of proposals as a national agenda, and there is little to draw from his small-ball legislative approach to suggest that he could succeed," the new article read. "Mr. Sanders is suddenly promising not just a few stars here and there, but the moon and a good part of the sun, from free college tuition paid for with giant tax hikes to a huge increase in government health care, which has made even liberal Democrats skeptical." That's definitely less glowing. We can put aside the "stars, moon, and sun" analogy for the moment—an awfully ornate (to say nothing of opinionated) touch for the *Times*' news pages. But while a shift from a positive valence to a negative one probably isn't a correctable error per se (indeed, depending on your persuasion, you might think the edits improve the piece's accuracy about Sanders's chances of accomplishing the things in his platform), it is a deeply important change for readers' understanding of the world around them.

The way a paper like the *Times* effects change in the world is generally not through lies but through emphasis and omission. Neither version of Sanders that the *Times* presents is *inaccurate*, but each nudges the world in a very different direction. Or, as Sullivan notes, "a reader who saw the piece when it first went up might come away with a very different sense of Mr. Sanders's legislative accomplishments than one who saw it hours later." She also notes that "the Sanders campaign shared the initial story on social media; it's hard to imagine it would have done that if the edited version had appeared first."

When a site changes the substance of its content so dramatically that those who shared it might regret it: This is a now-common way to scam. Sometimes it's making a Facebook group called, I don't know, "Parents of Golden Retrievers." You build up an audience and then start hawking bitcoin. Other versions would replace trafficked sites with malware. Others, like the *Times*, might not be scamming you, but they definitely aren't growing the reader's trust.

While defenders of stealth editing say readers won't care unless it's truly an error, here Sullivan notes an important way in which they might. If a puff piece becomes, suddenly and without notice, a pan, their understanding of the world—the very thing that they're sharing and commenting on—has broken. Michael Tackett, the editor of the Sanders piece, defended the tacit changes to Sullivan. "Tackett referred to 'the blessing and the curse of real-time capability,' and he said he made changes to developing stories every day." You might remind Tackett that having the capability doesn't mean it must be used.

Politically fraught drift and rot doesn't happen only in news. It can happen whenever the physical container of a piece of media is separable from the experience of the media. Andi Ortiz reports in *TheWrap* that Disney+, an internet video-streaming service, has scrubbed out a lone episode of *The Simpsons* in Hong Kong—one that contains a reference to the Tiananmen Square massacre. A single episode of the sitcom *Community* has been likewise scrubbed from the streaming services Netflix and Hulu. In it, a character wears dark makeup reminiscent of blackface. Tackett, the *Times* editor, called the internet's changeability a blessing and a curse; it's also one of the few things about the internet that doesn't change.

The medium is the malleability. The issue is who is doing the changing, and for what reason. The website NewsDiffs takes snapshots of the *Times* and other outlets' websites at regular intervals, and it allows users to compare one version to another. An account active on various social media platforms, Editing the Gray Lady, shows changes in *Times* headlines and subheadlines. It's not uncommon to see wars erupt between editors as a particular angle is changed and put back again and again. Tools like these need only exist in a world of unidirectional links and pages that can change on a whim. If news sites were harder to change once they published, Editing the Gray Lady would be awfully boring.

Nook, line, and sinker

But not all drift and rot is motivated by politics or money. Sometimes it's just stupid. The brick-and-mortar bookseller Barnes & Noble created Nook, their ebook reader, as an answer to Amazon's Kindle. In one Nook version of *War and Peace*, every instance of *kindle* had been changed to *Nook*. As in "to Nook a fire"—or as one astounded reader found: "It was as if a light had been Nookd in a carved and painted lantern." It seemed, at first, as if Barnes & Noble had reached into readers' ebook readers and awkwardly tried to remove references to Kindle. That's not what happened. Actually, it's likely that the publisher of that particular ebook had automatically found every instance of *kindle* and replaced it with *Nook* when making the new version for Barnes & Noble readers from its Amazon-specific one, according to reporting from *ArsTechnica*. That's not to say publishers couldn't—or wouldn't—fiddle with your ebook library. As Zittrain notes on his blog, Amazon once deleted a version of George Orwell's *1984* from its buyers' Kindles. Amazon founder Jeff Bezos ended up apologizing for the move, which he called "stupid, thoughtless, and painfully out of line with our principles." He didn't apologize for the obvious, innumerable, and painfully on-the-nose "Orwellian" jokes that followed.

There is precedent for digital artifacts to be disappeared. Zittrain, in his book *The Future of the Internet—And How to Stop It*, writes about an analogous case that played out in court. In 2004, TiVo, which allowed its users to record television shows and play them back later, sued EchoStar, a satellite TV distributor, for infringing on TiVo's patents—copying their functionality, in short. "A Texas jury found for TiVo," Zittrain writes. In 2006, "the court ordered EchoStar to kill the DVR functionality in products already owned by 'end users': millions of boxes which were already sitting in living rooms around the world with owners who might be using them at that very instant." You might think the shows you've recorded are, well, yours. But if you don't control the hardware and software on the machine that holds your stuff, your stuff isn't really yours. "Imagine sitting down to watch television on an EchoStar box," Zittrain says, "and instead finding that all your recorded shows had been zapped, along with the DVR functionality itself—killed by remote signal traceable to the stroke of a judge's quill in Marshall, Texas." Many of us don't need to imagine it. Occurrences like these have only become more common as, more and more, we cease to own the media we consume.

Sometimes it takes only the idea of a judge's quill. Tarpley Hitt, in a pair of 2021 pieces for *Gawker* (a website that no longer exists), reports that G/O Media—which runs *Gizmodo*, *Jalopnik*, *Deadspin*, the *A.V. Club*, *The Onion*, and *Jezebel*, among other sites—and *BuzzFeed* removed thousands of images from old posts. This wasn't a minor change. Hitt writes, "The *Onion*'s section 'News in Photos,' which typically features a joke headline and a photo, is now missing most of its photos." *BuzzFeed*, Hitt reports, removed photos to avoid legal copyright claims. G/O Media didn't make a similar announcement, but Hitt speculates that they, too, removed images to avoid the long arm of copyright law.

In 2021, a Twitter user with the handle @dox_gay discovered that several major news sites, including the *Washington Post* and *New York Magazine*, featured hardcore pornography in addition to the news. By way of explanation, Matthew Gault and Jason Koebler reported in *Motherboard* that a now-defunct video provider called Vidme had been bought by a porn site. News pages that had embedded a video from the erstwhile YouTube competitor now embedded pornography.

Those words that researchers use—*context drift* and *link rot*—feel inadequate these days to describe the chimeric face of the internet. In the days since I created my website, what we call a *website* has shifted. More and more, a website is actually a hodgepodge of texts, images, and other pieces of code, drawn from all over the web. A news article might contain ads, trackers, lists of links, embedded social media posts or videos—all pulled from other servers. The word *drift* implies a starting point: a place from which an object can move. But websites aren't static artifacts with initial positions. They are more like performances, a medium inherently ephemeral and unfixed.

When your browser requests a website from a server, it doesn't actually get a website. Rather, it gets the instructions, the code, for how to build the page. It's the same as the abstract notion of a piece of music—a Bach cantata, say—in which the ideal is never achievable; interpretations vary from musician to musician; meanings change from audience to audience. Similarly, different kinds of browsers produce different results, different personal software settings cause different interpretations, different contexts warp meanings. There are many musicians, all of whom must work together to perform a website. And, depending on the music (or website owner) in question, a different orchestra might be waiting in the wings, ready to swoop in and play a very different piece altogether. But we don't arrive at a concert and expect that the music will be the same each time. We'd be bored if it were. The ceaseless performance, the circus to which we all have front-row seats—this is what makes the internet, at its best, a joyful, human place to be.

Every performance ends, of course. Its joy is necessarily fleeting. When we close a tab, we close a curtain. When we pocket our phone, we empty a seat. The illusion of an effortless ongoingness, of a monument that needn't be maintained—this we must break. It isn't that everything ought to fade; it's that, online, everything will—unless we do something about it.

Fahrenheit 404

When I visited the headquarters of the Internet Archive, I saw, in the sanctuary of its old church building, dozens of statues of employees flanking the pews. Behind sat a server rack—a rectangular box of sleek black machines. The rack was taller than me, and wider too, though not by much. On each machine, points of light blinked in and out, visible even with the sun pouring in through the sanctuary's windows.

"Those are active servers," Mark Graham, the director of the Wayback Machine, told me, pride in his voice. "Those are live." Each light that blinked, he explained, represented some person's computer reading from the Internet Archive or writing to it. Each rack could hold several petabytes of data—each petabyte holds one thousand terabytes, or, to put it another way, one trillion novels. Graham measured the archive's growth to me in petabytes, more than a dozen in 2021 alone.

On either side of the dais, raised gold numbers announced pi (what a bunch of delightful nerds!), 200, 404, and 451. Graham quizzed me about the last three, and I could only recognize 200 and 404—internet-speak for a good response from a server and a "not found" error, respectively. I asked him what 451 meant, and, coyly, he told me it's the error code for "removed for legal reasons." It's hard to imagine numbers that better describe the archive's mission: 200—the Internet Archive—the last bulwark between 404, not found, and 451, removed by government actors.

Graham's zeal was understated but also obvious. The petabytes of data they collect are saving the digital world from decay. From the universe's cold entropy, we produce order; online, from the ever more chaotic morass of the web, we produce meaning. The people of the Internet Archive, memorialized as faux-terracotta soldiers in the sanctuary, are that meaning's keepers. It moved me. The public world ought to be kept, ought to be inscribed with as much force as we can muster. It ought to be stored in that true sense of the word—nurtured and replenished. And too often, it isn't. In the same way that Editing the Gray Lady would be boring if our internet were more stable, the Internet Archive wouldn't need to exist if the web's rot were more controlled.

There were once other models for networking together human knowledge, models that wouldn't have allowed for rot or drift. Ted Nelson coined the term *hypertext* in 1965, though its roots go back to 1960, to Nelson's first attempt at an experimental "writing system" in his first year as a Harvard graduate student. He wanted to create a "universal, democratic hypertext library that would help human life evolve into an entirely new form," Gary Wolf wrote for *Wired* in 1995. In the thirty years between Nelson coining the word *hypertext* and Wolf writing his piece, Nelson made a lot of promises but little progress. His project, his dream of joining hyperlinked pages together, called Xanadu, did not help human life evolve. It didn't even launch. It was, in tech parlance, *vaporware*: an idea for software, not the software itself. Xanadu, Wolf notes, "sucked Nelson and his intrepid band of true believers into what became the longest-running vaporware project in the history of computing."

What exactly was the dream of Xanadu? From the 1960s, fast-forward to the new millennium, and the project had not yet materialized, but Nelson continued to write his manifestos, describing what he wanted to create. "It has always been much more ambitious, proposing an entire form of literature where links do not break as versions change," Nelson wrote, comparing Xanadu to the web. In his vision, we could have an internet free of the "unidirectional link"—the way, now, one page points to a second, but the second can change at will.

In 2014, Nelson finally launched his prototype, OpenXanadu. Alex Hern at the *Guardian* describes OpenXanadu as "a simple document created using quoted sections from eight other works, including the King James Bible and the Wikipedia page on Steady State Theory." Half a century of development didn't produce much in the way of flash, it turns out.

The lesson here isn't that we should go invest in Xanadu. No, what Xanadu teaches us is that the history of technology brims with contingency. Xanadu, or something like it, could have been more than vaporware; it could have been the internet. So why wasn't it? Why did the web win out over something like Xanadu? For starters, it's less expensive, both computationally and in dollars and cents, to have an internet populated by unidirectional links. It's also more flexible. Finally, if you trust the others creating websites, why would you need Xanadu-style confirmation that the things they change aren't important?

That trust, though, has been eroded. It misleads—the abstracted, digital nature of the web—because at the core of link rot is the question of who ought to control access to the physical servers that actually comprise the net. In the case of some beloved, now-dead publications, it seems like the inhuman, invisible hand of the advertising market switched the servers off. In the case of the *Times*, it was the opaque desires of some editor. Sometimes, it's foreign governments. Other times, it's local judges. What's missing? The readers, the users: us, in other words.

Whether online or IRL, our media are driven by time toward entropy. We find agency in the choice to resist or succumb. The philosopher John Durham Peters writes in *The Marvelous Clouds* that "above all, media capture and fail to capture time, whose fleetingness is the most beautiful and difficult of all natural facts." In other words, the internet is at its most human when we acknowledge its ephemerality and make a choice to keep some parts and let others rot. If only the web were designed to help us remember that.

Shelf life

I wrote, confidently, that the Library of Alexandria burned to the ground. It's a story we're all familiar with: Humanity gathered all the world's knowledge in one place, only to lose it all in a vast apocalyptic blaze. But a more skeptical account of the myths surrounding the Great Library's destruction would note that not one of the alleged torchings carries much historical logic—or much evidence, for that matter. It's said that Julius Caesar burned the library in his fight for the city; more likely, the fire he ordered at the docks damaged some of its outbuildings, leaving the bulk of the library intact. Others claim a Roman civil war in the third century razed the library, but then what were the early Christians burning in the fourth and fifth centuries when, some say, they destroyed it not once but twice? And when the caliph Omar took the city in the seventh century, why did stories of his destruction of the library only emerge over five hundred years later?

The tidy, almost cinematic story of the library—a great wonder of the world laid low by the vainglory of generals and emperors, or the blinkered religion of small-minded zealots; by our human nature, in other words—tempts because it appeals to human vanity. It casts us as powerful, tragic figures, capable of creating a monument to our knowledge, and capable of destroying it just as completely. We want a myth like the Tower of Babel—another great human endeavor, destroyed and scattered by our own hubris. But, in truth, Heather Phillips, a librarian for the US courts, writes, "the fortunes of the Great Library waxed and waned with those of Alexandria itself. Much of its downfall was gradual, often bureaucratic, and by comparison to our cultural imaginings, somewhat petty." It's difficult to swallow that we lost the library slowly, out of disinterest. It's a harder story to tell that "neglect is the ultimate enemy of the archive," as Max Neely-Cohen writes.

We tell ourselves that the internet is a kind of modern Library of Alexandria. Maybe that's true. But if it is, then it's not because we face some impending cataclysm. It's because we're already in the long decline. And worse, we're walking into it unaware, whistling past the graveyard of the scrolls of Alexandria. "Over the past two decades," Neely-Cohen writes, "museums and archives have raised and spent billions of dollars to digitize their holdings, to say nothing of the countless individual citizen archivists painstakingly assembling digital collections on their own. Our hardware and software infrastructure is not built for this reality. It is tailored to the short term, without any concern for its long-term durability." We're rebuilding the Library of Alexandria, yes, both in its grandeur and in its long, whimpering decline.

CHAPTER 3
NEWS

The only living boy in Minneapolis

Paul Simon and Art Garfunkel claimed that you could gather all the news you need on the weather report, and, as a child, I was hard-pressed to disagree. Growing up, I loved the idea of the newspaper—the world, condensed down to ink and paper and brought to your doorstep. I loved the ritual of it, the way it pushed on a tender spot in my heart when my parents swapped sections over the breakfast table, alternating the front pages and the metro section. Or maybe it was the way they read to each other when they discovered something revelatory, my dad cursing the cowards in Congress or my mom relating the latest statehouse misadventures. I grew to love the fact of the newspaper too—actually reading it. There were the funnies, which I pretended to disdain. There were the articles, which I dutifully skimmed. But it was the weather that fascinated me.

Every day, I read the weather report from the *Minneapolis Star Tribune*—the then-name of my once-local paper—which was, most of the time, written by Paul Douglas, whose byline and author photo are seared into my brain. I would announce to my family what kind of jacket we ought to wear or if we needed an umbrella. I slotted myself into this morning rite, and it felt like belonging.

There's a picture of me sitting at one of the sidewalk tables outside the corner store in my childhood neighborhood, summer gathered all around in a haze of pollen off of which the golden light glints. In front of me is a bag with a half-eaten sandwich and an empty bottle of orange juice. I'm smiling at the camera, proud, the *Star Tribune* open in my hand. I can't be older than eleven. I was probably reading Douglas's weather report. Douglas wrote with humility about the uncertain future, but his words dictated so much about the day-to-day experience of my childhood. Would there be rain or shine, snow or ice? Would I go out and play, or would I be stuck indoors? A newspaper describes its city—its people, culture, and scene—and in so doing helps create the city's sense of place. What can be more central to that mission than the weather?

Minnesota's Twin Cities were blessed with two sizable daily papers: the *Minneapolis Star Tribune* and the *St. Paul Pioneer Press*. There was also an alternative weekly, *City Pages*. The *Star Tribune*, called either the *Strib* (affectionately) or the *Star and Sickle* (derisively), was *my* paper, the one my parents subscribed to, though the *Pioneer Press* would work in a pinch—as unfashionable but stolid as the city it described. I wouldn't discover *City Pages*, with its music reviews in the front, voice-y features in the middle, and salacious ads in the back, until high school. Once I did, I would make pilgrimages to the hip coffee shops that dotted Minneapolis to read it with my friends after class. I used to brag about the Twin Cities with the pride of a person who has never lived anywhere else. "We have three whole newspapers," I'd say, like, *look at how much ink this city is worth*.

In 2003, as President George W. Bush threatened to swerve the country into an Iraq-shaped ditch, I skipped classes and took to the streets. The very next day, over breakfast, I read through the *Strib*'s front-page story about the protests, the pictures of the crowd holding signs, all our mouths open in unison as we chanted slogans. Just like that, my marching took on a new importance. I was seen, made legible. I was *making* the news.

My parents still subscribe to the *Strib*, though they supplement with the *New York Times*. When I visit them, I pick up the paper of my childhood, now called the *Minnesota Star Tribune*, and feel its heft, lighter than I remember, almost gaunt. I wish I could say that it was only nostalgia filling out the stacked sections of local papers in my memory, but it isn't. *City Pages* shut down in 2020. Its website no longer functions, and its archives are now housed at the Hennepin County Library and the Minnesota Historical Society. The Pulitzer Prize–winning *Pioneer Press*, now owned by Alden Global Capital, has shed jobs and coverage areas, and what little remains of its dried husk is squeezed for ever more profits. As of 2006, 202 union staff worked at the *Pioneer Press*, Jay Bolter writes at *Racket*. By 2023, twenty-nine remained.

As Bob Collins writes for Minnesota Public Radio, "There's good money in destroying your local newspaper." Alden maintained an impressive 17 percent operating margin across its newspapers as of 2017, according to Ken Doctor at *Nieman Lab*. It accomplished these stratospheric profits, in part, by slashing head count—laying off reporters, in other words. And maybe this is obvious, but you can't cut reporters without diminishing the number and quality of articles a paper contains.

As the physical newspaper diminishes, so too does the city it describes. "When a newspaper dies in America, it is not simply that a commercial enterprise has failed; a sense of place has failed," Richard Rodriguez writes at *Harper's*. This isn't just speculation; researchers have found again and again that a city's civic energy, which demands a sense of place, really does slacken along with its local media.

Over a twenty-year period, "a decline in newspaper staff was associated with less competitive mayoral races and more incumbent-only mayoral races," write Jay Jennings and Meghan Rubado, academics at the University of Texas. Their data also suggested that "the decline in newspaper staffing reduced voter turnout." Worryingly, "on average, newspapers have cut nearly half of their staff since the early 2000's." More than staff cuts, newspaper closures can have dramatic effects on civic engagement in a city.

When the *Seattle Post-Intelligencer* and Denver's *Rocky Mountain News* shuttered in the late aughts, "civic engagement in both Seattle and Denver dropped significantly," writes Lee Shaker, an academic at Portland State University. He did not find similar drops across other cities without newspaper closures.

As the city's civic sense wastes, its residents turn away, cease bearing witness to their own communities. "The decline of local newspapers has contributed to the nationalization of American politics," researchers Joshua Darr, Matthew Hitt, and Johanna Dunaway write. "As local newspapers close, Americans rely more heavily on available national news or partisan heuristics to make political decisions." In other words, one consequence of the death of a city paper is the death of a city as a distinct civic entity. And the sense of place—a messy concept—matters not only to the health of democracy but also to our understandings of ourselves. What does it mean when your city loses its placeness? It means its residents have lost the part of themselves that was attached to it. It used to be, Rodriguez continues, that "the relationship between observer and observed was reciprocal: the newspaper described the city; the newspaper, in turn, was sustained by readers who were curious about the strangers that circumstance had placed proximate to them." Now, in this waning age of city newspapers, "we no longer imagine the newspaper as a city or the city as a newspaper."

The internet, most observers agree, caused the decline in local newspapers' fortunes. Much of a paper's journalism was subsidized by the advertising and subscription revenue generated by its classifieds and entertainment features. As ad dollars and reader eyeballs moved to sites like Craigslist, and as the limitations of physical paper competed with the inexpensive flexibility and rich media of the web, newspaper management failed to react quickly and lost revenue and subscribers. Online, while readership might, in some cases anyway, be up in aggregate—news websites don't suffer from a lack of traffic, after all—a vanishingly small proportion of causal news readers are willing to actually pay, especially at the local level.

In other words, people might very well want to read the news, but the business model that animated the industry has failed to keep pace with the market changes wrought by the internet.

A GeoCity on a hill

But the internet doesn't *need* to lack for a sense of place. In fact, the internet can open up new ways of understanding what a place even is. GeoCities, that now-defunct, once-beloved, free web hosting service for spectacularly ugly web pages—and one of the earliest nodes in the mass-market web—had the idea of place front and center; *city* was right there in its name. Its denizens called themselves *homesteaders*, a fine enough description of the people who tamed the early internet, tilled plots of arable land to grow *Grease 2* tribute pages, but the mixed metaphor rankles; one doesn't often think of homesteading in a city. Homesteading, after all, is conceived of as a solitary act, but GeoCities was a community. Its inhabitants swapped screen names for chat applications, linked to each other's pages, and wrote in each other's guestbooks—those early precursors to a comment section, where little more than a handle and a few words were permitted. I prefer to think of that time online as a metropolis.

Around the same time that the *Seattle Post-Intelligencer* and the *Rocky Mountain News* were being stripped for parts, the one-time Google rival, Yahoo, killed off GeoCities. Yahoo had bought the menagerie in 1999, a decade earlier, at which time it claimed 3.5 million users and 34 million pages. When Yahoo killed it, it deleted at least a terabyte of data. While it might not seem like it now, given the massive amounts of storage our phones and laptops now house, that's a lot of data. A single character of text might be a couple of bytes, depending on the encoding used. A terabyte is over a trillion bytes—roughly 500 billion characters, or 100 billion words, or one million good-length novels. GeoCities held more than text, of course; there were images too, and animated GIFs and tinny

MIDI songs that autoplayed whether you wanted them to or not. If a picture is worth a thousand words, surely a looping GIF of a stick figure peeing on another image (a common sight on GeoCities) would be worth ten thousand, at least. I didn't make my home on GeoCities—I built my first home online with Tripod, an early GeoCities competitor—but the metaphor worked for me too.

Cities, after all, are defined by their size, density, and governance. The early web I inhabited was expansive, especially when compared to my claustrophobia-inducing middle school experience, marked by bullying and loneliness. But paradoxically, the web was also bounded—a walled city whose streets could be walked and learned. The internet seemed to hold thousands of secret meanings from its thousands of users, all rubbing against each other, shoulder to shoulder, thick with signs and symbols that I could tease out with time and effort. With few friends (time) and something to prove (effort), I surfed my way across this new digital world, always looking up—up to that GeoCity on a hill, a beacon of community shining out across my dark adolescent world.

Online, a "sense of place" becomes less tangible, but it's no less real. The *New York Times*' website is a place. Reddit's website is a place. My old website remains a place, despite its abandonment. It's right there in the words we use: a site, a location, with an address. The comments on a recipe on the *Times*' cooking app, an unexpected bit of praise on Reddit, an email from an old friend about a post you wrote decades ago—digital social activities generate an almost civic experience. We are citizens of these sites, if we put in the effort. We feel pride in and ownership over where we live, digitally. So, when a beloved website dies on the internet, it is not simply that some code has been lost; a sense of place has been lost.

Weather or not

When I was a child, I collected the paper from our Minneapolis stoop and pawed through the (in-color!) weather report on

Sundays, and when I was in my early twenties, in my studio apartment in my small Ohio town, I collected the day's reading from my RSS reader and, cup of coffee in hand, warm light streaming in through the window above my desk, dug in, still checking the weather first. *The Awl*, a bloggy website active in the teens, contained Tom Scocca's "weather reviews," in which the author literally rated and reviewed the weather. On July 17, 2012, for example, Scocca gave the weather in New York City three stars out of five. "A solid peak-summer day, built on clean, baking heat," he wrote. "Late in the afternoon, as the lowering sun poured crosstown from the west, the dazzling glare and the blast of warmth felt united and directional; walking into the solar radiation was like walking into a strong headwind." These reviews, almost always from faraway New York City, let me feel like I was witness to—or even a part of—that thick mass of cosmopolitanism, rather than what I thought of as my provincial outpost in rural Ohio. The "place" to which I belonged, when I read *The Awl*, was a New York City of the mind—a city whose steel and concrete analogue I would find, later, that I disliked. But such is the power of the internet: It creates places with a shared culture, unlimited by the constraints of physicality.

Incredibly, the very forces unleashed by the internet, the same ones that toppled local papers, are now toppling the digital sense of place as well. Now, right now, as I'm writing this section, I found out that—finally, six years after it ceased publication—the website for *The Awl* has begun to flicker. *The Awl* defined the aspirations of my twenties. In the same way that the *Strib* and *City Pages* taught me new ways to see my city, *The Awl* taught me new ways to see the internet. I came to *The Awl* to read its coverage of the struggle between the internet I thought we had and the internet we actually have, which its editors called "The Content Wars."

The Content Wars chronicled the slow digestion of internet publishers by internet platforms. It's stupidly ironic that *The Awl* would eventually succumb to the technological forces it decried.

But progress must progress, or so I'm told. "Links from newspapers, from magazines, from blogs, Tumblrs and Instagram accounts; links from tweets, links in old emails, all are broken," the editors of the blog-like digital publishing experiment *Flaming Hydra* wrote in eulogy. "The original promise of hypertext was its ability to establish connections between ideas in disparate places and frames of reference. But nobody realized how fragile this system would prove until it began to shatter." The internet, the biggest mirror humankind ever crafted, reflects communities back to themselves, makes a user legible to themself and to the wider world. More than bad luck follows after it breaks.

Pixels by the quart

My first "real" job—that is, with benefits and a stable salary—came in 2013, when I helped build advertisements and infrastructure for those advertisements for the business news publication *Quartz*, which had launched only a few months before. Over the next few years, I would go on to work on *Quartz*'s product side, helping to build the website itself, and then, the very website I built would host my paid stories that carried my byline. As I write these words, *Quartz*'s corpse is just barely warm. It wasn't a dignified death. Initially launched under the umbrella of Atlantic Media—a kind of sister publication to the *Atlantic* magazine—it never made money, and was eventually sold to a Japanese financial data firm before being bought back by *Quartz*'s cofounder, who, a couple of years later, sold the site to G/O Media, a company known for the antics of its CEO, Jim Spanfeller. Notably, Spanfeller announced the acquisition in an email with the subject line "Quarts News."

In 2025, Zach Seward, *Quartz*'s former (at various times) owner, editor in chief, head of product, and cofounder, reported that G/O had "fired the few remaining writers at Quartz and sold the carcass to a Canadian firm that appears mostly interested in the email list." The site is still publishing what appears to be work from various freelancers, but I haven't had the heart to dig deeper, except to save my own articles from the rot that is no doubt coming for them. Anything beyond that would feel like grave robbing.

Quartz "tried literally everything in just ten years," Choire Sicha wrote for *New York Magazine*. It found chattering-class success in its initial decision to eschew a real home page. Instead, it placed articles in an "infinite scroll," with new articles appearing in, roughly, reverse chronological order—with some editorial guidance. (It would, eventually, try a home page.) It was designed to be shared. You were supposed to come in, as the lingo went, "through the side door." That is, you weren't expected to type "qz.com" into your browser—that's the front door. You were expected to find an article on some other platform, in an email newsletter, or from a search engine. If you clicked on that, you would be sent to an infinite scroll, containing every article *Quartz* had ever produced, with the one you wanted on top. Frictionlessly, just by scrolling, you would get the next article (and thus the next ad) right after you finished the first one. *Quartz* also experimented with newsletters, for which it received acclaim from media navel gazers and readers alike. It would follow the great "pivot to video"—that time in the late teens when it seemed like every company was laying off print reporters to focus on short- and medium-form video.

I wrote about media's pivot to video for *Quartz* in 2017. I argued that Facebook, which was then the source for nearly half of traffic to publications' websites, seemed to be inspiring the pivot, luring publishers with promises of money and eyeballs. The argument, as I understood it, was that publishers should "go where the readers are" and dedicate resources to places like Facebook in order to, hopefully, transform some portion of their users into your readers. "'Go where the readers are' is what publications whistle to themselves as they slink by the graveyard of their inflexible or unlucky brethren. 'Go where the readers are' is what they whisper to themselves as they quiver before the mercurial platform gods, waiting to find out if they'll be cherished or crushed," I wrote in a piece for *Quartz* from 2015. In retrospect, I'm surprised they let me publish it (and in such ornate prose!). That kind of willingness is rare in companies betting their future on the very trends I was criticizing. But maybe I shouldn't be surprised; Quartzians were never shy about dissent.

Despite my protestations, *Quartz* embodied the "go where the readers are" ethos; that was the entire point behind the lack of a home page at launch. And now it's dead. I don't mean this to gloat, but I was right about social media platforms' crushing embrace. Eventually, Facebook, after directing highwayfuls of traffic to publishers like *Quartz*, would direct those eyeballs elsewhere. Mostly, they'd try to keep them on their own platform. And in the lean times that followed, the media industry learned the hard way that borrowing traffic from Facebook worked only if Mark Zuckerberg agreed that news sites were worth sending traffic to. A city requires citizens. If all one has are commuters, it will lack the sense of place that defines the best kind of websites. *Quartz* was designed to capture commuters and, we hoped, turn them into residents. The gamble didn't pay off.

But perhaps this is too harsh. Toward the middle, *Quartz* had its newsletter business, which I know had its loyal fans. Toward the end, *Quartz* had a much-praised app, which its users often hailed as innovative. And, besides, who could turn down the wave of readers from Facebook et al. (but mostly Facebook) that powered so much of *Quartz*'s growth? In his book *Hamlet's BlackBerry*, the author William Powers calls the ideology, online, that more is more and more is better "digital maximalism." That ideology animated *Quartz*. And, in a different but related way, it animates today's digital platforms. For Facebook, Instagram, X (née Twitter), LinkedIn, and so on, more is always better—more friends, more likes, more posts, more connections; the users of those sites and services, too, are nudged to care about *more*. But the delight of a city is its finitude, its constraints, and the paradoxical abundance found within it—an encouragement to find deep connections rather than broad ones.

We can be loneliest in a crowd, after all. "I am so lonely I must subscribe to three papers," Richard Rodriguez writes, "the *Wall Street Journal*, the *New York Times*, the *San Francisco Chronicle*. I remark their thinness as I climb the stairs. The three together equal what I remember." I share Rodriguez's loneliness. In our platforms' quest for more, we have ended up with less. Perhaps it was time for *City Pages* or *The Awl* or *Quartz* to cease publishing; I don't know. But I do know we require what they represented: a collective sense of place, a city reflected back to us. These sites rotted away not because they couldn't find readers, not because we weren't enjoying their fruits, but because they couldn't find a business model. As the platforms and search engines take up greater and greater shares of advertising revenue, as they rework their sites to incorporate artificial intelligence summaries of the websites they once linked to, the web of distinct cities does indeed homogenize, flatten down, rot away.

Mourning edition

On my phone and on the watch to which it connects, I get the weather report every morning. It tells me—long before I fire up my laptop's inbox, long before I've even taken my first swig of water and started up my laborious coffee process—the weather. It knows my exact location and, thus, the exact weather I ought to receive. What it tells me is just for me and my family—just for the residents of my house on the outskirts of my small city in New York's Hudson Valley. I've also, in a fit of nostalgia, set it to tell me about the weather in Minneapolis, where my parents and siblings live; in New York City, where my colleagues complain unendingly about the mercurial climate; and in rural Wisconsin, where my grandmother had a cabin I loved as a child. I can control the weather report now—can be as practical or nostalgic as I decide to be. I used to love the agency and control that I felt this digital weather report offers. But more and more, it feels hollow, atomizing. Not because it isn't worth knowing what the weather is like where I am but because Paul Douglas's cheery prose has been replaced by numbers ringed with icons on my phone. Part of being a citizen entails knowing and caring about not just your conditions, but the conditions of your fellows too, and I appreciated Douglas's tacit insistence that we could locate these civic bonds even in the shared experience of the weather.

These days, I get up before the sun rises. I sit on my porch, facing east just off my kitchen, and I watch the sky lighten behind the maples and oaks and lone birch by the little stream that divides my land from my neighbors'. This is where I read my local paper, the *Daily Freeman*, which still comes with a weather report—though it's shorter than Douglas's ever was. "Breezy this afternoon," it announces, "with a shower." Reading the *Freeman*'s prediction, I feel the loss of the kind of community that could support a multiparagraph description of the weather. Still, when I read the *Freeman*'s pages, my eyes have no choice but to flit across the full weather report, to learn about the highs and lows across the Hudson Valley and beyond. A table of data gives me the weather in Cleveland and Cairo, Tokyo and Tripoli.

Today, a blue bird with an orange breast kept me company on my porch while I read the weather. He distracted me admirably from a task that, more and more, inspires only loneliness. Today, the sun will rise, and the clouds will cover it, but only for a while—and only ever imperfectly. The men with their mowers will come and make the air smell like summer. My daughter and I will marvel at the bean plants in our little garden that feasted on this weekend's showers and yesterday's sun. In town, the patio at the pizza place will be full of young families, and the couples on their bicycles will wear T-shirts and shorts. At some point, I'll look up the bird on my phone and discover its name (eastern bluebird). I'll look up from my phone just as the clouds clear out and the sun startles a smile onto my face. I'll rate the weather five out of five. None of this will bring *The Awl* back to life, or the *Pioneer Press* back to health, or my childhood back to me—and it shouldn't. But Tom Scocca still reviews the weather in his newsletter, and my town, twenty-three thousand strong, has both the *Daily Freeman* and the *Kingston Wire* to make us known to ourselves. I may be lonely, but at least I'm not alone in my loneliness.

Part 2

. . . AND SO WILL YOU.

Have the poets left in the garment a place for a patch to be patched by me?

—Antarah ibn Shaddad al-Absi
(trans. F. E. Johnson)

CHAPTER 4
DEATH

Hello, world!

I'm a sucker for ritual. Small ones: Every morning, I write in my journal. Big ones: Every year, at Christmas, I re-create a nineteenth-century Anglican celebration called the Nine Lessons and Carols with my friends and family. Work-related ones: Every time I learn a new computer language, I first use it to write a program that prints "Hello, world!" onto the screen.

That last practice originated in 1967, with the Basic Combined Programming Language, called BCPL, according to the *Jargon File*, a long-running repository of hacker slang. Now, Wikipedia features dozens of tutorials with "Hello, world!" implementations. I, myself, have written "Hello, world!" what seems like countless times as I've learned new frameworks and new languages in my career, and there's something almost wholesome about those thirteen characters coming alive on a terminal or browser window—the program's exclamation point never ceases to bring a smile to my face.

This ritual also links my informal, journeyman education in software to a long chain of master coders, and that fact, more than the exclamation point, probably explains my fascination with the practice. After all, in many languages, "Hello, world!" is a terrible introduction. In Python, for example, the time to Hello, World is laughably short—you simply write, *print("Hello, world!")*—while in others, such as C++, you would need to include a library of code that can handle outputting text, a "main" function—an entry point into the code—and a bunch of boilerplate that instructs the code how to move the characters of "Hello, world!" out onto the screen. There's little reason, these days, that we learn "Hello, world!"—except for the ritual of the thing. For all our obsession with progress, software engineers tend to be remarkably nostalgic. That nostalgia, an almost hoarding impulse, means we squirrel away data like nuts in the fall. We keep every bit and byte we can. It's better to have it and not need it than to need it and not have it, and besides, digital storage keeps getting smaller and cheaper.

But ritual, nostalgia, and an impulse to hoard, which can power wholesome rituals like "Hello, world!," also can be misused. We remember what we ritualize, and bad rituals perpetuate bad outcomes. A misogynistic ritual reproduces a misogynistic world, for example. Take the so-called Lenna image, used as a standard test in parts of computer science. Drawn from a 1972 *Playboy* centerfold, the model Lena Forsén wears a somewhat absurd hat with purple feathers on it, and she looks back at the viewer over her bare shoulder, a Mona Lisa smile on her lips. Cropped just below her neck, there is no nudity, but there's a kind of flirtatiousness to the image, a charged quality. "The story of the image," Jennifer Ding writes at *The Pudding*, "follows the story of the internet," and, spoiler alert, it's not a happy story.

Ding traces the use of the image across the web and in academic papers, starting in 1972 and ending in 2021. These days, the image lives on without Forsén's consent. "I retired from modeling a long time ago," she says in the documentary *Losing Lena*. "It's time I retired from tech, too." The stakes of the image's continued use (over 240 instances of it appeared in Ding's data set in 2021) might, at first blush, seem small, but as Ding writes, "When one of the only women this well referenced, respected, and remembered in your field is known for a nude photo that was taken of her and is now used without her consent, it inevitably shapes the perception of the position of women in tech and the value of our contributions." Not everything *should* last. Not everything is designed to be kept. Sometimes what you produce is just ephemera. Sometimes what you produce should be consigned to the dustbin.

Think piece be with you

Speaking of things that ought to be binned: On that first website I made back in middle school, a two-paragraph essay I wrote still lingers, annoyingly. I must have been reading some degrowth manifesto, because in it, I'm offering a now-embarrassing Malthusian-flavored take about food and population growth. "The world, as we know it anyway, seems to only get smaller and smaller, yet our population only gets bigger and bigger," I write, blaming this growth for our environmental collapse. "Why produce more than we use? We have been producing more than we need ever since the Fertile Crescent, so our population has been exploding ever since than [*sic*] too. If we want to stop destroying the earth, we've got to stop growing. We've got to stop making more than we need." I read these lines from my essay to a friend, a poet. They had just enjoyed an edible and were playing *Mario Kart* on their handheld gaming console while we talked. Still, they noticed something in this argument that I hadn't (poets, man). The idea that, driven by a capitalist, consumerist culture, we make more than we need rhymes with the argument I advance in this book: Driven by a technocapitalist, digital culture, we keep more than we need. As a result, in both cases, we suffer from a kind of ecological collapse—one in our physical environment, the other in our attentional one.

My partner, the writer Galen Beebe, once said in an essay that "my obsessions don't change, but maybe they are different now that I am different." This passage has become a joke between us. She catches me rewatching a television show I loved as a teen, and she says "my obsessions don't change." I see her playing with American Girl dolls with our five-year-old daughter, and I tell her, "Maybe they are different now that you are different." It's true, it seems, that my obsessions haven't changed, but I have. I no longer think we should restrict growth to prevent ecological collapse. I've become, instead, something of a digital ascetic: You can't take it with you, and trying to hold on, miserly, can cause nothing but trouble. Digital death isn't only inevitable; it's necessary.

At the core of a workable society is forgetting what should be forgotten and remembering what should be remembered. My faux pas with a friend fades away, but my good deed sticks with them—I hope, anyway. In truth, most of what we do, most of the time, deserves to be forgotten. Our juvenile actions, our ill-considered utterances—these are not meant for an audience beyond the present. If we went through life remembering every slight and awkwardness, we would end up stuck. "Time," to quote *The Awl*'s John Hermann, "is a privacy setting." Or at least it used to be—at least it should be. More than a privacy setting, time heals all wounds, including the social ones.

"A word is dead / When it is said," Emily Dickinson wrote—a fact the theorist Walter Ong would make much hay over. In a purely oral world, Ong notes in his book *Orality and Literacy*, words are "occurrences, events." A spoken word "exists only when it is going out of existence." Ong contrasts this with the written word, which can be "detached from its author"—which lives on whenever a reader sees it. A rebuke online carries more power because I feel slighted each time I see it, whereas a verbal reprimand might be forgotten as the vibrations that carry it are swallowed up by time. When monks copied and recopied their monasteries' books—doodling in the margins and illuminating the author's words—writing was the exception rather than the rule. Today, most of my communication with many of the people I'm in community with, I'd hazard, takes place in written form. I text many friends more often than I speak with them. As a result, even our quotidian discourse is searchable.

To win an argument, a friend recently exhumed an old prediction I made—a bad one. He was right and I was wrong, I'll admit it here, but those words represented a different me, one I would have preferred stayed buried.

I've grown obsessed, in my own fumbling spirituality, with practices of redemption. In Judaism, the faithful burn their leavened bread, a symbolic burning of pride itself. In confession, Catholics are offered a chance to speak their sins aloud, receive penance, and are absolved. In some strains of Protestantism, baptism washes away spiritual grime, leaving behind only an inward grace. I've made my way to my own practice of amends to those I've harmed. I've learned the hard way that defects of character must be handed over to make way for the sacred. The internet as presently imagined seems to have forgotten this. Platforms don't forget and rarely forgive.

Bánh education

My college appears in the news every few years due to the activism—or perceived activism—of its students. In one particularly absurd dust-up, in 2015, a small piece in the school's paper of record, the *Oberlin Review*, noted that the dining halls were butchering food from some Asian cuisines. The "bánh mì" didn't use the ingredients that define the dish, for example. The "sushi" included "undercooked rice" and no fresh fish. But the piece namechecked "cultural appropriation" in slamming the terrible interpretations of these cuisines. That, apparently, was enough to inspire national attention. "Students at Lena Dunham's college offended by lack of fried chicken," the *New York Post* blared. After that, the race was off. The *New York Times*, the *Atlantic*, and innumerable mocking tweets soon followed. "Bánh mì is cultural appropriation" was the (not ever spoken by the students) line, and, removed from the complaints of the students in question, it does sound ridiculous.

The web isn't just a time machine; it's also a transporter. It allows its users to move through both time and space. In researching the bánh mì incident, I was able to travel back to Ohio, to pull up the *Review*'s decade-old original article, just as I was able to find the dated, mocking *Post* piece and the resulting mean-spirited social media posts. I never worked for the *Review*. I worked as a copyeditor for the *Grape*, the alternative paper. It came out biweekly and included a seminude centerfold and a section called "Bad Habits." The style guide, which I helped to write, needed entries for sexual acts that don't look at all nice in print. The *Review*'s website serves the Oberlin community. It also, apparently, serves the editors and reporters of the national press on the "college students are too woke" beat. Thank God the *Grape* didn't have a website when I was working there. Thank God that era has been allowed to stay put in Ohio and fade, gracefully enough, into my past.

"A person's life is, of course, comprised of all its moments, and a person comprised of all his or her experiences taken together," Elizabeth Bruenig writes at the *New Republic*. "But the internet, with its instantaneous responses and global reach, tends to make the passage of phases harder than ever. Each of us, it increasingly seems, is one selfie-stick-snap or ill-considered tweet away from being permanently crystallized in a single moment of our immaturity, like insects preserved in amber." And it doesn't even need to be the stench of real immaturity; it's enough that the internet gets even the slightest whiff of it.

Even more, the algorithms we rely on—Google Search, for example—take into account our past activities. Spotify uses what I listened to before to guide my future listening. TikTok knows what I've watched and, crucially, as I reported for the *Wall Street Journal*, leverages how long I dwell on each video to decide what to serve me next. The web traps our utterances in amber, yes, and more and more, it traps every one of our digital actions, no matter how small or automatic.

Rot can be dangerous. But rot also sits at the heart of cultivation.

In its proper place, when it enables fertility in the field, we call it compost. To compost is to control rot, to harness it for the purpose of growing more and better. Rot in and of itself isn't the enemy; it can even be a friend. When we embalm our past selves—trap ourselves in amber, as Bruenig puts it—we are refusing to add our old ways of being to the garden's bounty.

Of course, online, often the choice to let something rot isn't really ours. Not too long ago, I set up my tweets to self-destruct after a few months, under the logic that, to paraphrase Robert Hass, a tweet is mostly an elegy to what it used to signify. The conversation that surrounded it had already moved on; why should my contribution to a months-old discourse be remembered? But this process required a half hour, a dozen steps, and mysteriously stopped working several months ago. Social media networks might not want our words to be ephemeral. Large language models, which require tremendous quantities of human writing, can gorge themselves on all those old posts. And even before large language models, our old posts fed the platforms' content machines. It might cost money to host my words on X, but they make a good deal more in advertising revenue and investor dollars.

The issue with rot is that we ought to be the ones managing it, and we ought to control where it's allowed to thrive. There is no utopian past to point to, when humans had perfect control over their utterances. Before the internet, an author could be held accountable for a passage in a book they wrote. The difference is that few people wrote even a single book, and those books would in all likelihood eventually go out of print and cease to be relevant. Now our words—many novels long—linger on servers we don't own, ready to be resuscitated by a single search.

Maggots have their place in the compost pile, not the kitchen. All of this is to say that we should, in fact, allow websites to change or even go dark. When a site's creators and netizens are ready to move on, they should be allowed to. To change the metaphor:

Cities aren't—and shouldn't be—static places. The closure of a site like *The Awl* vexes precisely because it seems senseless, outside the power of the people who love it. By analogy, when a community decides how to use its neighborhood, which buildings to tear down and which to renovate, the city is stronger as a result. But when outside forces rot the city as a whole and from within, even the idea of the city weakens, both as a place and as a civic institution.

A community might still run roughshod over some of its individuals' aims for their artifacts, I don't doubt. That's the hard work of community building—work required both on- and offline. But though perfect agency might be an illusion, communitarian agency enables a richer relationship to our digital selves and stuff than the offerings we have now. In short, today's web often couples the immediacy and urgency of orality with the permanence of literacy—but capriciously. Neither individuals nor communities get to decide when to expend resources to enshrine our speech, to lock it away as though it were a prized possession.

Online, there are models of a more compostable web. On Signal, correspondents can set their messages to self-destruct. On Snapchat, images can disappear after being viewed. Even on Wikipedia, that pillar of the web, the site's editors change its entries in a constant flow, ensuring that they're up-to-date and accurate. They remove entire pages, large swaths of text every day. There's a way to see what was changed—the rarely viewed and sometimes difficult-to-parse "history" of an entry—and there's some friction there: layers of difficulty between you and a past that's no longer relevant. In other words, Wikipedia manages its rot through its public "talk pages" and history, which detail the vast civic energies that power and shape its removals. That's opposed to the way algorithms on our social media platforms pull old utterances and display them for you without even being asked.

Offline, as we rush forward through time, the past gets further away, harder to access. Online, all too often, the past remains stuck with us. When we enter a query into a search engine, we are, as often as not, turning on a time machine—one that imperfectly captures what we once said, since so much of the past has rotted away.

Dewey's decimals

If we feel our access to the promised magic of a digital world is precarious, contingent, and contested, that's because it is. The historian Lara Putnam writes that after two Toronto papers' archives were digitized, "citations of these two newspapers in history dissertations increased tenfold." But, she writes, "citations of their undigitized contemporaries remained steady or fell." Digitized records of old newspapers are a boon to researchers wondering what life was really like back then, but as Kaspar Beelen and his coauthors ask in a 2023 paper on Victorian-era papers in England, "Whose past is actually captured by digital collections? Whose voices or perspectives are underrepresented or even completely absent?"

Newspapers in the "mainstream"—a euphemistic word if ever there was one—of American life have long been digitized. Meanwhile, in 1973, the Moorland-Spingarn Research Center at Howard University took over the archive of the National Newspaper Publishers Association, "a trade association of Black-owned community publications," Kenichi Serino writes for PBS, but it wasn't until 2021 that they received funding to digitize all those old newspaper articles. At the Library of Congress, it wasn't until 2022 that a wide swath of African American newspapers was added to its "Chronicling America" project. But "for some titles," the library notes, "only a single issue is extant or available." What we choose to keep, what we choose to toss—this choice is suffused by the attitudes of the powerful. And as Beelen et al. note, even if a researcher were able to sample every newspaper from the time, "such a corpus would still reflect the hierarchies of power that governed Victorian society; reproducing its social prejudices and stereotypes." History is written by the victors—and in their voices too.

When I attended conservatory, studying baroque performance practice, I would fumble my way through the music library's stacks almost every day, hoping to stumble across a book that could teach me something living about all those dead composers whose music I was learning. Sometimes, it would be a piece of music I'd discover. Other times, a treatise that referenced the recorder. Every once in a while, a connection between the deep reservoirs of meaning in the wider historical world and the music that was soaked in it. I worked out of the second floor of the conservatory library, the airy windows casting cheer on the study carrels, etched, as often as not, with little messages from studiers long ago. When my mind revolted against the demand that I keep reading, I'd stretch my legs and loosen my brain by walking through the rows upon rows of books and scores, looking for serendipity—serendipity being so often the parent of insight. Who knew that Bach wrote a vanishingly small number of tremendous sonatas for recorder and harpsichord? I didn't—at least not till I went looking for Buxtehude's sonatas and ended up with Bach's instead. My serendipitous discovery, alas, was made possible only by a reverent institutionalism not bestowed on everyone.

"Innovations that cultivate serendipity—such as the Dewey Decimal System, by whose graces a trip into the stacks for one book often leads to a different, more salient discovery—must, almost by definition, be plagued by arbitrariness," Bill Wasik writes in an essay for the *New York Times Magazine*. But arbitrariness, and thus serendipity, does not drape itself across the topography of human experience evenly. That is, arbitrariness, in this power-mad world, isn't actually arbitrary. Indeed, where it mountains up—like most things—has much to do with the human institutions that shape our lives. Wasik gives the example of a book about the Mariposa Battalion, a California militia active in the nineteenth century at the heart of the California genocide—the ethnic cleansing of Indigenous people who lived in the Golden State as it was becoming a state. Shelve that book with others about the gold rush, "and it will acquire a very different set of neighbors than if it's classified as a book about the Battalion's victims," Wasik writes. There is no neutral way to organize information about the human world because humans always take a side.

For the researcher, this presents a key problem: They don't know what they don't know. The serendipity of discovery in the stacks cannot be left uncomplicated by all the other serendipities they could have found if only the stacks were arranged differently, if only the world were arranged differently. To guard against all those unknown unknowns, researchers have spun up all sorts of cultural practices: They conduct countless interviews, engage in deep research, searching exhaustively. But this kind of analog research, the friction of it, often involves dramatic amounts of inefficiency. As a reporter, I will regularly talk to an academic for several hours to get a single usable quote, a lone paper that makes its way into the story, or that one insight that changes the course of my reporting. I'll often spend hours digging through some archive or other, looking for a specific piece of information—reading, processing, thinking about a million and one seemingly irrelevant strands. And it is exactly this kind of seeming inefficiency that leads to true understanding. So yes, on the one hand, digitization means that I can "process" data wherever and with ease. On the other, to process means to read and think about what you're reading, to paraphrase Putnam. When historians "don't stay around long enough to be inconvenienced, insulted, or instructed, the quality of their analysis suffers," she writes. "The forced contextualization that made traditional historical research inefficient looks, on reflection, like a significant contributor to knowledge production." But that inefficiency is being magicked away.

"Algorithms fetch for us," Putnam writes, "doing away with the need for intermediaries like brick-and-mortar stores (if you are Amazon.com) or regional expertise, bibliographies, and immersive reading (if you are a historian)." She terms this digital discovery *disintermediation*. But I'm not so sure about the prefix *dis-* there. Even as digital technologies cast down the walls between us and the world we seek to understand, new ones spring up, ones that can almost be invisible. It's as much *dis-* as *remediation*. We have been given a wilder, stranger kind of intermediary: the information systems structured by the internet. As Putnam herself writes about digital technologies, they are making "new realms of connection visible, new kinds of questions answerable. At the same time, the new topography of information has systematic blind spots. It opens shortcuts that enable ignorance as well as knowledge."

It's not just historians who suffer these systemic failures. The relative dearth of, say, non-English-language texts online, the emphasis on the lived experiences of the digitally savvy and the literate, the subtle influence of what our platforms have chosen to save rather than toss—together these threaten to mislead our contemporary research just as much as our historical. "Technology has exploded the scope and speed of discovery," Putnam writes. "But our ability to read accurately the sources we find, and evaluate their significance, cannot magically accelerate apace." This is our problem as much as it is our historians'. As we rush to replace one set of intermediaries (librarians, say) with another (a large language model masquerading as a librarian), we are calling up new and different mediators, not doing away with them. The need will always be there for something to mediate the world and our minds; the world's unruliness demands it.

Despite all its vastness, the internet we can find, search, and hold in our minds is a vanishing fraction, a rounding error, of the human experience. But precisely because it is so vast, we can trick ourselves into believing it is a good-enough proxy for humanity. When I was working at the MIT Media Lab, I worked on a project called Electome, which had programmatic access to the full firehose of Twitter posts. My colleagues created complicated machine learning algorithms that sifted through millions of tweets. One striking finding following the 2016 election of Donald

Trump—a surprise to many journalists—was that if you overlaid whom users followed and were followed by on top of what they talked about, you could see how disconnected many "verified" journalists were from the topics and users who supported Donald Trump. Hillary Clinton's ascension to the presidency, to many in the media, seemed like a fait accompli, but how much of that perception was driven by their placement in a rich environment devoid of Trumpian voices?

"Digital search offers release from place-based research practices that have been central to our discipline's epistemology and ethics alike," Putnam continues. Place-based research might not be a panacea for our unknown unknowns (to paraphrase, of all people, Donald Rumsfeld), but despite the jokes about *New York Times* journalists parachuting into diners in Middle America, there's efficacy in unplugging from the digital world's false fullness. At least it can offer a reminder that your milieux does not sum to the world's.

Mystical opportunity

"It is past time to put aside other business and to turn our energies to the remystification of virtually everything," Marilynne Robinson writes at the *New York Times*. She's suggesting, all the way back in 1987, that we have become too entranced by our own materialism and that our obsession with what we can measure has blinkered us to greater truths about the world. "I am as firmly persuaded as anyone that things can be explained, or usefully described," she writes. "But I am convinced that our methods at this point are crude, and premature, and that they lead us to ignore such experience or intuition as cannot be accommodated by them." This is a theme Robinson returns to again and again in her work. In *The Givenness of Things*, she tells us that we live in a culture of naturalism in which we pay "exclusive attention to the reality that can be tested by scientists." In *The Death of Adam*, she sees that "we have reached a point where cosmology permits us to say that everything might in fact be made of nothing, so we cling desperately to the idea that something is real and necessary, and we have chosen, oddly enough, competition and market forces, taking refuge from the wild epic of cosmic ontogeny by hiding our head in a ledger."

Comprised as it is of ones and zeros, the digital world might seem to be the final form of the ledger, a place where mysticism cannot take root. But we live online—we humans, stubbornly non-digital. Despite our fundamental humanness and the richness it implies, online, we are stuck with a kind of naturalism, to use Robinson's terms: one in which we pay exclusive attention to the reality we can horde. But the taproot of mysticism is our finitude—and the question of what comes after it. To accept that our stuff, which includes our bodies, might sicken and die—there begins a belief in something greater. To remystify the web, we must first acknowledge that we can die there.

As is so often the case with juvenilia, my old essay was at once right and wrong. I was right that, these days, we produce at a functionally infinite scale, that we make and store much more than we need. I was wrong that this state of affairs is new, a solvable problem, or even a problem to solve. What matters is choice over what we choose to store and consideration over how we access it. Our newfound ability to recall ever more of what we've warehoused, powerful though it may be, will not, at long last, set us free from death. The human condition will have the last laugh; eventually, even my anime site will fade, just as I will. Keeping it or my fragile body alive takes scarce resources, including attention, time, and will—and those things will fail, in the end.

The liberation theologians argue that the kingdom of heaven must be constructed here on earth. As Archbishop Oscar Romero said, if the gospel is "a message about eternity and the hereafter," then "it is also a message about the here and now." It is "a message about the demands of this earth, about the fulfillment of our duties in this world." I take him to mean that the care with which we treat the fickle sacks of water and meat that comprise our selves and our neighbors matters. Ignoring either our inevitable deaths or our capacity for living unjustly—this is what Scripture means when it warns against false idols. Considering the online human condition, I see a culture that has allowed new erasures and injustices to flourish even as it has turned away from that one, great absolute truth: We will die. And in turning away, we have turned a technology that could liberate into one that constrains. In turning away, we have turned a mysterious, awe-inspiring world into one that holds only the meanest facsimile of what it means to be human.

CHAPTER 5
ATTENTION

Just one more byte

It's possible that *to store* is entirely the wrong verb. Maybe it ought to be *to attend*, to give our digital lives the same attention we are called to give our physical ones. *Attention*, as a noun, comes to us through Latin and French, and, from those languages, carries the sense of *to stretch toward*, as though my brain, pinned behind my skull, might reach out, feel the tight restriction of muscle as it grasps at the world. But, by and large, attention words point inward, not out. We attract it and call it. We can gather it like wheat in a field or draw it up like water from a well. Always, though, attention moves from the world into us—except when it's to be minted and spent. "Pay attention," we say, as though it is a commodity, bought, sold, and traded by brokers.

The digital world both beleaguers and demands our attention. The quality of our focus and the metaphors we surround it with have diminished even as we're asked to spend more of it. I've felt it, lately, in the very medium of the book, which rests ever more uneasily in my hands, though I've published two, including this one. I've read, I would guess, a couple thousand books in my life. It's harder now, harder than ever. And as it's grown harder, it's grown more meaningful, those acts of attentional resistance, as straightforward as reading a book for pleasure—a refutation of market logic, a proletarian revolution in even the word *attention*, not something I pay for and sell on an open market but, instead, something I stretch toward.

Goodreads, the Amazon-owned website that lets you track your reading and review and comment on your and others' books, holds my reading history since the start of the COVID-19 pandemic, but I attend to that history too. I have a spreadsheet, a simple one, not a Microsoft Excel file but a text file marked by commas and new lines, readable by both machine and human, where I record each book's metadata (author, publisher, publication year, etc.) and tag them all with one or more of two dozen-odd tags I've invented, my own personal ontology of what I value: *family is complicated* or *the bawdy politic*. Every few months, I take stock of my reading, which is how I know I'm reading more poetry than ever, even as the act of reading it grows more difficult.

"Whole strata of reality are lost to us at the speed at which we live, our ability to perceive them is lost," Garth Greenwell writes in his novel *Small Rain*, "and maybe that's the value of poetry, there are aspects of the world that are only visible at the frequency of certain poems." Greenwell advances an argument best advanced in literature, in fiction or poetry, buttressed as it is by the logic of allusion. This logic stands in stark contrast to the totalizing, seductive logic of instrumentalism, which would have us see everything, every act and artifact, as for something else. Reading under instrumentalism is pointless at worst and a form of self-betterment at best—merely a way to refill the coffers of attention before spending it again. Greenwell argues for poetry, for reading, for all those many varied noninstrumental activities that demand this "unmixed attention," to use the philosopher Simone Weil's language, which are a species of prayer. I mean this in the same way Weil does—in the best and truest sense of the word: prayer not as an attempt to change the world, but as a chance to attune ourselves to a greater will, one that asks of us, *Be more human*.

In his poem "We," Joshua Bennett writes that "the money of the mind is attention, maybe." I love that *maybe*. After all, taken at face value, and in the unflattering utilitarian light of our internet age, Bennett's speaker's assertion seems correct: Money makes the world goes round—as it has since it was invented, I don't doubt. And, of course, there's never enough of it, even as it takes on ever more importance. We need the stuff, and the world seems determined to strip it from us.

And then, there's the flip side of the coin, the other ways of thinking about money. In the Christian tradition, for example, money and access to it could be highly regulated. Christians often forbade usury—charging interest on loans—and even now, in some circles, the practice is frowned upon. Early Christians used to venerate ascetics, hermits, and the like, and many Christian sects still take vows of poverty. Jesus did say, in multiple Gospels, that it would be easier for a camel to pass through the eye of a needle than for a rich man to get into heaven. Despite some contemporary Christians' efforts to ignore or twist Christ's meaning there, I take him and his early followers at their word: more money, more problems—at least spiritually speaking.

The push and pull between these two understandings of money—as life-giving and essential on the one hand and as spiritual poison on the other—makes it a fraught, if relevant metaphor for today's ever more instrumentalizing attentional markets. But Bennett's poem continues. "Sustained attention is how we approach / a flesh and blood experience of the Divine," he writes, hearkening back, yet again, to Weil.

Speaking of someone with enough money to cause problems, the erstwhile cryptocurrency titan Sam Bankman-Fried claimed not to read. "I don't want to say no book is ever worth reading, but I actually do believe something pretty close to that," he said, as reported in the *Washington Post*. "If you wrote a book, you fucked up, and it should have been a six-paragraph blog post." Columnist Molly Roberts responded, "You're supposed to read not in spite of the digressions and diversions that stand between you and the denouement, but because of them. . . . What you come out of a book with isn't always supposed to be instrumental at all." She draws a line from Bankman-Fried's reading life (or lack thereof) to his moral philosophy (or, again, lack thereof), the same one that led him to be convicted of crime after crime. Those actions "would only count as wrong if the ultimate outcome of them was likely to be wrong, too," she writes. "What happened along the way, just like with those books not worth reading, didn't appear to carry much water." But even this argument can be twisted into an instrumentalist one: Reading fulsomely can make one more moral. Perhaps, this argument goes, if Bankman-Fried had read Greenwell, he wouldn't have defrauded (allegedly) quite so many people. I don't know. I doubt it. I do know that reading to attune oneself to the human buzz of the human world brings me a deep, resonant joy, and maybe Bankman-Fried can cultivate that joy, wherever he finds himself next.

Books don't make readers more moral, but reading—like so many other acts of attention, ones that fill us with that deep joy—can exercise the kinds of attention that enable moral agency. As the philosopher Silvia Caprioglio Panizza writes in her book *The Ethics of Attention*, "Attention shapes us and our world. It constitutes the background upon which deliberation, choice, and action occur." What we choose to attend to, how fully we're able to attend to it, whether it's our choice to attend to it at all—these constrain us. What reading offers is a chance to glue back together the broken shards of our attention, a chance to make moral choices with care. But books aren't unique in this regard; resonant attention can be found even online—though it's become a remarkable exception rather than a rule.

Information superhighway to hell

In the 1990s, while working for Tripod, the very hosting service that still holds my terrible first website, Ethan Zuckerman invented the pop-up advertisement. Tripod, after casting about for a winning business model, settled on a tried-and-true way to make money off its rack after rack of user-generated content: Sell ads. Zuckerman and his colleagues wrote algorithms that read the content of the websites their users made and served ads based on their inferred interests. Like, if you're on a website about car maintenance, you might be interested in buying car parts. The system wasn't perfect. As Zuckerman recalls at the *Atlantic*, "A major car company freaked out that they'd bought a banner ad on a page that celebrated anal sex." Enter the pop-up ad, which seemed to solve Tripod's problem: Users get their ads, and companies have a little bit of remove from unsavory content. "I wrote the code to launch the window and run an ad in it," Zuckerman writes. "I'm sorry. Our intentions were good."

I tried, recently, to read a story on my local newspaper's website. As I scrolled through it, ads for goods I had already bought popped into existence, animating wildly. Some entered from the bottom, some played videos, others covered the text I was there to read. This experience repeats itself enough that it has ceased to mean much of anything in the same way repeating a word turns it nonsensical. We cannot read local news without being subjected to a flickering mess of boxes aiming to modify our behavior. And we have largely acclimated to this state of affairs, which seems as worrisome as the advertising itself. I don't blame Zuckerman, personally, for this hellacious reading experience, nor do I blame him for the eight-lane superhighway we hurtled down to get there. But I do blame the intention, which always contained an impossible contradiction: *attend to this article* on the one hand, and *attend to this ad* on the other.

Moral panic at the disco

It seems obvious to me that when I read a book, especially the kind of book that enables deep attention, I am engaged in a different kind of activity than when I read my local newspaper's website, though they share the verb *to read*. It seems similarly obvious that our attentions, and thus our selves, are casualties in that verb's degradation. But otherwise intelligent people often minimize this complaint—that the changing media environment is changing us for the worse. Plato, Postman, and plenty of media theorists whose names don't start with *P* have predicted sweeping changes to culture from writing, television, and more. Reasonable people often respond to concerns about the internet more broadly (or rapacious corporations hell-bent on algomagicking away our attention, more specifically) with the ever more in vogue rejoinder of "Sounds like a moral panic." In this reading, Plato's *Phaedrus*, skeptical of the utility of writing, simply shows "how all contemporary tech criticism is similarly misguided and short-sighted. Plato was wrong about writing, thus contemporary critics who adopt the same pattern of analysis are likewise wrong about whatever novel technology they happen to be complaining about," L. M. Sacasas writes in his newsletter, *The Convivial Society*. Describing criticism as a moral panic, like any kind of cynicism, is a cheap way to sound wise, but the thing is, Plato and Postman were, by and large, right.

That's not to say we oughtn't be humble as we analyze the media environment. "We should always operate from a stance of extreme skepticism that we live in a particularly important human moment," Freddie deBoer writes, calling for a Copernican revolution in culture writing. I find this hard to argue with: Every pundit thinks we're facing the most important election, every young person thinks they invented sex, and every media critic thinks, *This'll be the tech that changes everything*. I take deBoer's warning, though, as an opportunity to remember the past, not write off the present. Plato said that writing would "implant forgetfulness" in our souls. Was he wrong? Neil Postman wrote in *Amusing Ourselves to Death* that television would force entertainment to the fore of all our human activities, even politics and news. Was he wrong? Though I'm no Plato, and certainly no Postman, I say our digital lives—our avatars and artifacts alike—are purposely manipulated to favor breadth of attention over depth. Am I wrong?

Of course, my uncomplicated and unsurprising assertion that I'm right and others are wrong is complicated, surprisingly enough, by the rightness of my imagined critics. Just as Plato was correct, so too were the legions of people who saw the power of the technology of writing. Today, just as we who panic about digital technologies' choke hold on our attention are correct, those who see its potential are also correct—as are those who argue that all the panicking has unsavory precedents. These critics' illustrations of the in-hindsight silliness of moral panics, too, I find hard to argue with; there's a cottage industry of finding excellent quotations from all manner of long-dead thinkers about the dangers of some media technology or other, quotations that now seem bizarrely quaint. Letter writing, book reading, radio listening, and telephone calling: just four horsemen of societal calamity. My favorite comes from John Kellogg, who, in 1882, wrote that "the novel-devotee is as much a slave as the opium-eater or the inebriate."

The journalist Chris Hayes offers a useful way of thinking about these moral panics in his book *The Sirens' Call*. He highlights the frenzy over comic books in the middle of the twentieth century. "In the end," Hayes writes, "it turned out that comic books weren't worth the worry, which is why the panic looks silly in retrospect." But he contrasts this panic with similar hand-wringing over the introduction of tobacco. He cites England's King James I (of the King James Bible fame), who, all the way back in 1604, called cigarettes "lothsome to the eye, hatefull to the Nose, harmefull to the braine, daungerous to the Lungs, and in the blacke stinking fume thereof, neerest resembling the horrible Stigian smoke of the pit that is bottomelesse." This description would sound over the top, if you forget he's talking about such an unhealthy product.

Hayes offers a question: "Is the development of a global, ubiquitous, chronically connected social media world more like comic books or cigarettes?" I take Hayes to mean, in part, that maybe a *moral panic* is just a *worry* about something you think we oughtn't worry about, in much the same way that a *think piece* is just an *essay* you find small or a *trailer home* is just a *tiny house* for a poor person. Panic might be in order; it's just hard to sort that out in the moment.

I'd add to this that, sometimes, something like a moral panic can succeed, and we rarely call a *social movement* a *moral panic* when it does. Panic over children losing fingers in factories led to rules that prevented child labor. Panic over the state of our slaughterhouses in the wake of Upton Sinclair's depiction in *The Jungle* led to food safety laws. Panic over the steep decline of birds from chemicals like DDT after Rachel Carson's *Silent Spring* led to environmental regulations. We only call a social movement that argues that some technology might need amelioration a moral panic when it's not actually something that's causing problems.

Postman didn't really just bemoan the rise of television; he also argued that we needed to see "both the gains and the losses, the benefits and the burdens. Only then would we be able to judge soundly and wisely," as Sacasas summarizes. Children and their nimble fingers produced goods that kept costs down; unregulated slaughterhouses democratized meat eating; DDT killed mosquitoes, which carry myriad horrible diseases. We simply decided, as a culture, that these gains were not worth the losses. I'd suggest that it would be more intellectually honest if the people who cried "moral panic" would, instead, argue their case on the merits. Say why you think attention-harvesting technologies' benefits outweigh the costs; don't simply scoff at those who worry the losses outweigh the gains.

We can blunt a technology's problems—and in so doing let benefits flower without annihilating the soil—by spooling up cultural practices around it. Kellogg, who compared fiction to opium, might have been more correct during his own time than in ours, as norms around how often and when we can read a book are more developed now than they were at the height of the novel craze. And concerns and social pressure of the sort Kellogg espoused undoubtedly played a part in developing those norms. Any technology used without guardrails, without having been metabolized by culture and considered against competing technologies and practices, can be dangerous. My parents, by way of example, banned television on weeknights, though I could watch cartoons on weekend mornings. My parents weren't the only ones to enforce this rule, and the market took note: There were, when I was a child, simply fewer cartoons to watch on weekdays than on weekends. But digital technologies are new. Their impact ought to be measured in decades, not years—and yet we have to live with them here and now.

We don't have time to languidly develop our norms; we need them yesterday. Digital technologies exploit our shared attention in the absence of countervailing forces that would preserve it: This is the issue. The recent flood of books, articles, and documentaries, which some call instantiations of a moral panic, I would call a natural and essential immune response to any new technology.

Technologies—the internet, viral social networks, the even more virulent form of content-generating artificial intelligence, which the internet and social media unlocked—change and evolve, and they change us, demanding that we adapt. We still don't know how we will adapt to the cacophony our digital tools have unleashed, though we can guess. As *to read* changes from a verb that describes an attention-generating activity to an attention-sapping one, something that was numinous becomes instrumentalized, subject to an ever more utilitarian way of thinking. We all become a little more like Bankman-Fried.

Some interlocutor might point out that this particular complaint—that we are reflecting, more and more, a coldly utilitarian world—can be found in the past as well. "Even as a schoolboy I took intense delight in Shakespeare," Charles Darwin wrote in his autobiography. "But now for many years I cannot endure to read a line of poetry: I have tried lately to read Shakespeare, and found it so intolerably dull that it nauseated me." In the end, Darwin writes, "my mind seems to have become a kind of machine for grinding general laws out of large collections of facts," and when it comes to Shakespeare, "the loss of these tastes is a loss of happiness." It seems trading truth for beauty, beauty for truth, isn't new; people have complained since the cold clockwork of modernity was first wound up that it is ticking away the mystery and mysticism of a more human past. That makes it no less painful when this swap comes, newly, for something we love. Keats may have promised otherwise in his poem "Ode on a Grecian Urn" ("Beauty is truth, truth beauty," etc.), but in an age where AI algorithms optimize our attention into a commodity, his romantic vision seems less a promise than a prayer.

Sometimes, a difference in degree can be so vast that it becomes a difference in kind. Or, in other words, bringing poetry to bear on the problem is as useless as bringing a knife to our attentional gunfight. Greenwell's narrator, a poetry professor, alone in his hospital room as he is for most of the novel, vents his frustration about the inability to marshal arguments for poetry in the face of his students' instrumental skepticism. "But it was difficult to defend the alternative, to justify it in terms of outcomes and deliverables, costs and benefits." Poetry, he says, "was indefensible by that logic, its value lay in demonstrating the possibility of other logics, other relationships to value, I mean other ways to live." In other words, reading is not for attention, though it can recharge it. Reading is not for empathy, though it can inspire it. Reading is not for education, though it can give you one. Reading is for reading. And in reading, we find new ways to read—which is to say, to be.

Wild goose chase

The poet Mary Oliver died in 2019. She remains one of the few poets whose work I can cite and reliably expect to strike a chord of memory—at least among my cohort. Her accessible verses, which often contain a kind of rousing call to broadly defined mindfulness, still circulate widely, especially on social media. In the poem I see the most, "The Summer Day," her speaker demands, "Tell me, what is it you plan to do / with your one wild and precious life?" These lines' poeticism flows from the surprising word *wild*, its rhythmic partner *and precious*, and the well-placed line break. Coupled with the poem's sentiment and its directness, this passage became an instant favorite, an almost-meme. In the days following Oliver's death, my Instagram feed nearly burst with pictures of this poem. Even now, not a month goes by where I don't see a screenshot or two of her work. This poem, like so much of her work, implores readers to find refuge and meaning in the natural world, yet I read it most (and have even shared it myself) on a digital platform designed to capture and abuse our attention—a delightful little irony that I think Oliver would have appreciated.

But the accessibility that turned her into a meme has its discontents, alas; "certain critics liked to trash Oliver as unsophisticated," Franklin Foer writes at the *Atlantic*. Even buying that her poems lack sophistication wouldn't necessarily mean they lack value. That oft-quoted demand about your wild and precious life, for example, boldly and baldly stated as it may be, encapsulates the stakes of attention. Isn't that line "a road map for how to live—which is also to say, how to die?" Steven Petrow asks at the *New York Times*. I think it is, or at least I think it's a road map for what I believe about attention. Even Weil's belief that capacious attention is a kind of prayer echoes in Oliver's poem. "I don't know exactly what a prayer is. / I do know how to pay attention," she writes, "how to fall down / into the grass, how to kneel down in the grass, / how to be idle and blessed." In these lines, Oliver undermines the idea that her speaker doesn't know how to pray. She kneels in the grass, paying attention, and receives idleness—a blessing—in exchange. It seems Oliver knows exactly how to pray: by attending to the world around her.

Her speaker's urgent question from "The Summer Day" has become our question as we've repeated it on social media.

Answering it demands that we reckon with our finitude, with that most central fact of how we spend our time. Attention, our "endless and proper work," as Oliver puts it in a different poem, implies mortality. Internet platforms might not want to encourage this line of questioning, given that they rely on our choice to keep wringing out our minds long after the last drops of attention have run out. Taken together, Oliver's work represents a kind of *memento mori*, a Latin phrase and spiritual practice that means "remember your death." As Foer writes, "If the mind peels away distraction, it inevitably thinks about death." I wonder, sometimes, if this truth, which is again and again found in Oliver's work—that attention inevitably draws the mind toward death—doesn't explain the lethean pleasures of TikTok. After all, as Oliver writes in an essay, "Attention is the beginning of devotion." Who wants to be devoted to our own deaths?

During the COVID-19 pandemic, I read about a nun, Sister Theresa Aletheia Noble, who collected skulls in order to help remember death. "We try to suppress the thought of death, or escape it, or run away from it because we think that's where we'll find happiness," she told Ruth Graham at the *New York Times*. "But it's actually in facing the darkest realities of life that we find light in them." I first ran across Sister Aletheia's death practice on Twitter, where she was posting bite-size meditations on death, complete with skull emojis and Bible quotes. The incongruity—the same mismatch as Oliver's poems finding their home online—delighted me. Maybe the best place to attend to death is in on platforms that encourage us to forget it. Or, as Oliver writes in her poem "When Death Comes," "When it's over, I want to say: all my life / I was a bride married to amazement." To find the world amazing is to pay close attention to it, including to its fleetingness.

Sometimes it feels like I'm building a levee of books, a wall meant to hold back the rising panic that builds from time to time, demanding, "What, exactly, did you accomplish today?" Reading might be a joy, but it's an irksome, fickle joy that cricks my neck and humbles my pride and that I often avoid, mostly with computer programs—or *apps*, as they're now called, a diminutive word I detest, which obfuscates the truth that we are all carrying computers programmed to do things—on that glass-and-metal rectangle that is never out of reach. More than a joy, reading is also my last, best defense against the cold waters of instrumentalism, because unlike nearly every other one of my activities, reading captures my full attention while accomplishing more or less nothing measurable except the passage of time.

At work, I get to produce knowledge from time to time. That knowledge is even good and valuable, I truly believe. In measuring my time at work, I try to pick a useful metric. Not, say, lines of code I've written, but the quality of facts that my code has uncovered. It works most of the time. But if I think my life is full because I produce measurable things, I end up with a lot of things and very little life. So, reading. Ink on paper. The weight of a book. Time spent with body and breath, my mind both present and not. The moment when, chapter finished, I lift my head and sigh and am in the world once again and everything and nothing has changed.

That spreadsheet to which I attend, the one I store, remains an inefficient way to explore what I've read. But that inefficiency is generative in the same noninstrumental way reading itself is generative. It would be easier to house my reading life in Jeff Bezos's digital platform, more useful. There's a delightful lack of friction there, a slipperiness, to browsing Goodreads. Mark a book as "Want to read," purchase it with a click or two, and it arrives at your doorstep in a day or so. Tag your books in the right ways, and see more like it; let its algorithm do the hard, annoying work of deciding what you ought to read next. My spreadsheet, on the other hand, rubs against my will—a kind of friction. I hope it's a generative one. It's exactly that kind of friction, after all, that sparks a fire.

CHAPTER 6
FRICTION

Magic

Back in my middle school library's computer lab, I sat in front of one of their boxy machines and taught myself to speak internet on its no-frills, built-in text editor. I had to wait dozens of seconds for the modem to connect to the net. I'd tap my feet and squirm until I heard the telltale *screetch* and *bewip* sounds, oscillating in increasing dissonance.

These days, I'm a fluent coder, and when I'm not, I have artificial intelligence chatbots to help. I'm always connected to the web, if not on my laptop, then on my smartphone. The only sound the internet makes is the jolting hum of the vibrations from that rounded rectangle when yet another new email is delivered. It's easy, now, to take for granted my movement through the digital world, easy to forget the feeling I had then, in that computer lab—like this was the age of miracles. "The internet is paradigmatic magic," writer and critic Virginia Heffernan writes in her book *Magic and Loss*. "It turns experiences from the material world that used to be densely physical—involving licking stamps, say, or winding clocks or driving in cars to shopping centers—into frictionless, weightless, and fantastic abstractions." I loved the kind of sorcery I had discovered back then. I could conjure structure from thought. I could animate text with a single line of code. I could impress my will out onto the world, and someone—or tens or hundreds or thousands of someones—could see what I had created. All I had to do was type the right words in the right order. *Abracadabra.*

Heffernan calls it frictionless, the way the "densely physical" world becomes effortlessly traversable. But the internet of the early aughts—my internet—was not frictionless at all. Certainly not when compared with the glossy, corporatized web of today. My desires needed to be wrapped in HTML, in annoying, indented bits of only sometimes comprehensible code. As my wishes made their way out into the world, they left traces of themselves: those modem sounds and the *clicks* and *clacks* of the mechanical keyboards that echoed across the empty computer lab. The file-transfer process took too long, and at home, when siblings or parents were on the phone, there were long, chafing delays to even get on the web. The internet was constantly rubbing against my intentions as I pushed them out.

Friction online behaves metaphorically in the same way that friction offline behaves literally. Speaking digitally, I will go further on a website that requires two clicks to sign up than on a website that requires twenty—unless I'm willing to exert myself. And if I push an idea up against the rough side of an app's constraints, it'll lose some edges, just as if I pushed a block of wood against sandpaper. A thought longer than 280 characters? X will trim that. A reaction that isn't one of five emoji reactions on Facebook? Try something else.

That's not to say we don't commune in digital space with far less effort than before. We're living in the golden age of "user interface design"—of patterns that lubricate our web experiences, of digital products that go down smooth. We are nudged, gently, into funnels that encourage us to navigate or read or buy the things we're supposed to. And down we go. Even more, there are sites and applications that allow me to post whatever I want to populations several orders of magnitude larger than my 2001 website could have ever hoped to reach. The slim brick of metal and plastic in my pocket allows more digital life than I could have dreamed of in those early days.

We've gained so much in this frictionless world, but we've lost something too. Recently, a friend who was newly in love wanted to send his paramour a hand-picked batch of songs. When he was growing up in the aughts, he would have used a bulky computer to "rip" the music off compact discs, then he would have ordered the files for the songs on his computer and "burned" them back onto another CD. He would have decorated the CD with markers and, maybe, written liner notes for the case. Now, he makes a playlist on Spotify—a subscription service with a massive catalog of music. There is less friction—less physicality—these days when we want to make a mix tape. Putting aside the need for physical CDs and CD players (or tapes and tape decks), even the metaphors are physical—ripping and burning. On Spotify, you just add songs to a list. "It doesn't feel like a *thing*," he said, italicizing the words with his voice. He ended up deciding to make a website of sorts, hosted on Google Slides, where he would embed videos of each song along with liner notes. Not a mix tape, but at least more frictional. Online, it's easy living. But that means we are alienated from the mechanics, from the thingness, of the internet.

The web's "fantastic abstractions," as Heffernan calls them, often mask the internet's messy physicality, which must exist *somewhere*. In truth, the internet remains resolutely physical, made of copper, silicon, plastic, and steel. Despite its vaporous name, "the cloud" requires power and water in immense quantities. "A single data center can consume the equivalent electricity of 50,000 homes," Steven Gonzalez Monserrate reports at *The MIT Press Reader*. Collectively, they "devour more energy than some nation-states." Powering them, Monserrate writes, pumps more carbon into the atmosphere than the airline industry.

To cool the humming, sweating machines that store our ephemera, data centers drink in billions of gallons of water. Google's data centers, some of the most efficient in the world, consumed

6.1 billion gallons in 2023, according to a Google environmental report. A family of four uses about 150,000 gallons of water a year, according to the Environmental Protection Agency, which means Google alone is consuming as much water as over 40,000 families in a year. And these tangible ecological costs are growing, driven in large part by the ubiquity of artificial intelligence. Microsoft, whose data centers house OpenAI's models like ChatGPT, wants to "revive the shuttered Three Mile Island nuclear plant" in order to power its AI offerings, Tom Warren reports at *The Verge*. The company's own sustainability report notes that it is "not yet on track" to meet its goal of reclaiming more water than it uses by 2030. This failure is driven, the report notes, by the generative AI boom it helped to husband.

Giant, expensive buildings, windowless and nondescript, greedy for water and electricity, make up the metaphor, the abstraction, that we call the cloud. Onto it we have layered further abstractions, obfuscating the internet's underlying physicality. First *code*, then *companies*, and, finally, *users*. These ideas—code, companies, users—intangible as all ideas are, can seem agentless, just the rules of the game. But flesh-and-blood humans wrote the pieces of code that comprise the internet, decided which letters to imbue with the power to move electrical signals just so, and those signals, in turn, have the power to move our thoughts. And flesh-and-blood humans, summed together, bundled the code into products that give structure to the web. And flesh-and-blood humans stand behind every user whose clicks and swipes interact with those companies' products.

The move from a frictional web to a frictionless one is mirrored by the move from a human internet, one contingent on our fellow's flesh and blood, to an abstract one, one populated by corporations and institutions. Each enables the other in a cycle of ever-increasing alienation from the materiality at the base of it all.

Consider the doorknob

The dream of a frictionless digital world traces its history to early thinking about the computer itself. As Sherry Turkle writes in the twentieth anniversary edition of her influential 1980s ethnography of early computer users, *The Second Self*, many engineers and designers believed that "there was no need for a user to ever address a machine's underlying mechanism. In their view, computer users should be liberated from having to think about the machine at all." In his 1998 book *The Invisible Computer*, the influential designer Donald Norman excitedly predicted the vanishing computer. Media theorists Jay David Bolter and Diane Gromala summarized Norman's view in their 2003 book, *Windows and Mirrors*. "We'll continue to use computers, of course, but they will disappear from view as they are absorbed inside 'information appliances,'" they wrote. "We won't care much that there is a digital processor inside these appliances; we will care that each appliance accomplishes a task for us."

Bolter and Gromala point out that Norman has been, in part, vindicated. Today, we use objects that are really computers all the time: cars, phones, watches, thermostats, pressure cookers, Wi-Fi-enabled toasters. Computers have colonized the world, but we don't often think of them as computers at all. They're more like our gut bacteria—essential, though essentially ignored.

Norman illuminated his position with a physical metaphor: "A door has an interface—the doorknob and other hardware—but we should not have to think of ourselves using the interface to the door," he wrote. "We simply think about ourselves as going through the door or closing or opening the door." Among those who create our technological world, Norman's point is gospel. Digital artist Olia Lialina, in a talk titled "Once Again, the Doorknob," quips, "I really don't know any mantra that has been quoted more often in interface design circles." In her essay "Turing Complete

User," Lialina recounts that when the iPad was announced, Apple's release notes proclaimed, "We believe that technology is at its very best when it is invisible, when you are conscious only of what you are doing, not the device you are doing it with." Of course, thinking about a doorknob (a device) also means thinking about friction, the pull of hand against handle, of mechanism against mechanism. The future, if *Star Trek* or the local supermarket is any guide, is a place in which doors spring open as you approach. But the knob, and the friction it entails, can be useful: We think about the physical facts of doors all the time, as when a child is sulking in their room or when partners want their privacy. The knob allows its users control, where a lack of a knob allows its users to be controlled.

The ideology Norman espoused—one that encourages frictionlessness—doesn't apply just to the computer. It has taken root in the designs of the applications hosted on those computers, which include, first among them, the internet. Lialina notes that when Adobe announced Dreamweaver, which let designers create a website without coding, the character in the advertisement said, "I have more time to do what I like most: being creative." Lialina is having none of it. "The message from Adobe is clear," she writes. "The less you think about source code, scripts, links and the web itself, the more creative you are as a web designer. What a lie."

Large language models, powered by artificial intelligence, reduce the friction between thought and action online even more. Frictionlessness, raised brick by brick over the past decades of digital building, now pokes at the clouds; like in the story of the Tower of Babel in Genesis, we are coming close, dangerously close, perhaps, to the heavens. A user can simply ask a chatbot for what they want and receive code that works—for the most part, anyway. With some models, the chatbot will even execute the code for them.

The sturdiness of an object corresponds to the amount of friction it yields. A sturdy door is heavy, pulling just a bit against the ground, clicking satisfyingly into its jam. A sturdy computer program takes a little work to make it work, has a learning curve, a place to grow into. A sturdy website is embedded in a thick web of links, and pulling it free would tear it from its context. But our tech companies are sanding away the friction between us and our avatars and artifacts, between us and the sturdiness of the web. It might seem like a contradiction to note that the web's sturdiness is matched only by its fragility—but only until you remember that living things (you and me and everyone we know) are also both sturdy and fragile in the exact same way.

God at Chili's

Stretching from adolescence to now, I've been engaged in a project of experimentation with religion and spirituality, an attempt to generate a connection to the divine and transcendent. I'm practiced enough to observe my brain trying to wrest control from my body: As I wrestle through my prayers in the morning, I can watch as I convert the experience of flesh—of uncomfortable knees and struggling breaths—into one of mind. I hear myself protest that I cannot possibly find God in the body, unclean as it is, full of gurgles, burps, and farts, hair in strange places, and smells that defy easy categorization. But even as I struggle with my body, I've come to believe that I cheapen the ineffable when I squeeze it into a shape my brain can hold. Mystery might find purchase, instead, in the body.

There's a long, storied history to the Christian emphasis on flesh and blood. One of the religion's holiest rituals, the sacrament of Communion, in which Christians eat the body of Christ, reflects this centrality of the messy, human fact of the body—the locus of humanity's constraint, sin, and suffering. We don't chew his divinity. We gnaw on a body. "Yet the Catholic insistence that one eats God's body ('broken and chewed,' to cite Aquinas) not as symbol but as literal enzymatic and chemical fact—as metabolic feat—helps make sense of the gospels' otherwise baffling obsession with broken and suffering bodies," Ellen Wayland-Smith writes in her book *The Science of Last Things*. "There is no more visceral way to experience the fact of our shared flesh—to understand that our participation in the other is literal, not figurative—than to come face-to-face with the physical vulnerability of another's body." The Communion bread and wine, in other words, remind us that our body mediates *both* the material and spiritual world; that it is holy.

Saint Thomas Aquinas, a thirteenth-century "doctor of the church," whose philosophy shaped much of the Roman Catholic Church's doctrine, argued in his book *Summa Theologica* that the body exists even in heaven, every part—including hair and nails. Even "the entrails will rise again in the body even as the other members: and they will be filled not with vile superfluities but with goodly humors," he writes. I find Aquinas's insistence almost comical, his cataloging of which body parts, exactly, will accompany us to heaven charmingly credulous. And yet.

When we locate the divine in the lived experience of a body, we are implying something radical about what we are called to do. "Again and again," Wayland-Smith continues, "the historical Jesus of Nazareth scandalized his community by provoking precisely this gut-level reaction of outrage and disgust at our broken oneness. He walked around Galilee and Judea touching any and every person who might be considered low or unclean: rotten flesh, empty bellies, blasted eyes, dirty feet, bleeding prostitutes." In the strands of Christian thought that I am pulling on, Jesus located divinity precisely in the form of the most "unclean" among us. Christ ministered to them and asked his followers to do the same. The struggle to heal the sick, protect the poor, and recenter the outcasts—this extends logically from the sacredness of the body, even and especially in its brokenness.

In the sitcom *The Office*, the hapless boss of the paper company Dunder Mifflin holds an award ceremony for his workers in a local Chili's restaurant. The receptionist, Pam, wins an award—called a "Dundie"—and as she tipsily accepts it, she says, "Finally, I want to thank God because God gave me this Dundie, and I feel God in this Chili's tonight." I think we're supposed to laugh, though not that unkindly, at Pam and her winning, earnest declaration. But as in my reading of Aquinas, I've always found this line both funny and profound: If God can be found—felt even—in a Chili's, where else might the divine be hiding? Stuck in my house during the pandemic, I attended a Quaker meeting on Zoom. At first, I found this digitized version of a once-potent ritual—worship with Friends—ineffectual, but before long, sitting on my uncomfortable dining room chair, my body began to ache in the familiar way of the meetinghouse. Before long, I heard the communal silence just as I do in services, my breath leaking through the noise-canceling headphones. In Chili's, in the Zoom room, on the cross and in the tomb—wherever our bodies go, there too goes God.

The Gospel of John opens with a long description of incarnation—a deity taking human form. "In the beginning," John tells us, "was the Word, and the Word was with God, and the Word was God." Thirteen verses later, we learn how, through Jesus, "the Word was made flesh, and dwelt among us" (KJV). Flesh and blood—this is how we experience the divine. If a more secular metaphor is required, Robert Hass writes in his poem "Meditation at Lagunitas" that "there are moments when the body is as numinous / as words." The lesson I want to draw from Aquinas, Communion, and Christ is that our bodies, precisely because they constrain and fail—precisely because they are mortal—hold our souls.

Friction, I've found, often marks the spot where a numen makes its home, which makes sense when you remember that friction is a reminder of our limitations, of our bodies that push or pull or kick or drag but always, in the end, fail.

Dualism

These days, though, we mistrust the body—and the friction it generates as it passes through the world. I blame Plato. "The body is a source of endless trouble," Plato has Socrates say in *Phaedo*. "It fills us with loves and lusts and fears and fancies and idols and every sort of folly." Fast-forward some nineteen hundred years from Plato, and we have René Descartes claiming to do away with all knowledge gleaned from the body, since the senses so often mislead. Descartes splits the world into *mind* and *body*, two distinct and separable kinds of things, with the body being subordinate to the mind. The trouble is that when you devalue the body, it's easy to miss the tenuousness and fragility of the physical world that our minds—and websites—inhabit. Easy, that is, until something goes wrong. While I was writing this book, I slipped on a walk and dislocated my left kneecap, spraining a couple of ligaments as it ripped free of its housing. I had never really thought much about my kneecap—why would I?—but when it went, so too did my ability to write for a month. Under the kind of dualism espoused by Descartes, my writing, a product of my mind, shouldn't be so affected by my kneecap, part of my body. But it was.

Just as our bodies are never more front-of-mind than when they break, the cloud doesn't become substantial until something goes wrong. In 2017, an engineer at Amazon Web Services, the biggest cloud computing provider, entered a typo while performing routine maintenance. That engineer took down a vast swath of the internet for several hours and cost companies in the S&P 500 at least $150 million, according to the *Wall Street Journal*. There was also the Fastly snafu, the Howard Street Tunnel fire—more outages than I can count. We might wonder what exactly is up there in the cloud when AWS or Fastly has an outage. Most of the time, most of us are comfortable letting it remain, well, cloudy. But things do break. All the time. Not just the cloud, but the machines that fill it. Not just our PCs and phones, but the interface of websites we visit. Bolter and Gromala write in *Windows and Mirrors* that "no interface can be or should be perfectly transparent, because the interface will break at some time." Media like books, movies, and the internet mediate (it's right there in the name!) the viewer and the ideas being viewed; in other words, they are "interfaces" between us and the world around us. What Bolter and Gromala call *transparent*, I call *frictionless*; there are plenty of ways to describe the web's subordination of the body in favor of the mind. "The internet casts light onto many things, but rarely on itself," philosopher John Durham Peters writes in *The Marvelous Clouds*. "Like all media, it comes with a built-in cloaking device." Shrouded as it is, it's difficult to think about the internet's physical form. It's hard to pierce the cloud and even see the internet as having a form, even though that form, in fact, holds my digital life, which is increasingly just my life, which means, increasingly, the separation of my life from its body rankles.

"According to conventional views, the human relationship to technical things is too obvious to merit serious reflection," the philosopher of technology Langdon Winner writes. He points out that, in this conventional telling, "once things have been made, we interact with them on occasion to achieve specific purposes. One picks up a tool, uses it, and puts it down." But, he writes, "technologies are not merely aids to human activity, but also powerful forces acting to reshape that activity and its meaning." Winner's "conventional" view is eerily similar to the transparency gospel or to the illusion of frictionlessness. *Don't think about the doorknob; think about going through the door*. Or, *Don't think about the website; think about what you're doing with it*. There's money to be made and power to be had controlling media. Social media platforms, which craft the interfaces between us and our friends, extract a toll with every visit. For this reason alone, it's worth thinking about the sites and services we use. But too often, we sleepwalk through our digital lives, content to offload the work of determining how we came to the content we came to—a kind of attitude that Winner calls, memorably, *technological somnambulism*. Ironically, for me, what woke me up was the proximity of the medium of the internet to the longest sleep of all: death.

CHAPTER 7
ALGORITHMS

The hard disk jockey

In 1995's *Clueless*, a perfect adaptation of Jane Austen's *Emma*, the whiny tones of Radiohead play while the protagonist, Cher, puts on her makeup. "Yuck, the maudlin music of the university station," she proclaims. "Wah, wah, wah! What is it about college and crybaby music?" A fantastic question.

In high school, driving around, I used to demand that my friends and I listen to the University of Minnesota's radio station, Radio K, which you could find only on AM radio—what a novelty! We discovered all manner of bands on Radio K, mostly of the *wah, wah, wah* crybaby music variety, which seemed so sophisticated at the time. After classes, my friends and I would take the bus to Cheapo or Electric Fetus—two of Minneapolis's premier record stores—and paw through the CDs and tapes, looking for more songs by the artists we heard on Radio K.

We had many problems, my friends and I. None seemed quite so pressing as the construction of ourselves through the art we liked. Our question—what should be the next music we listen to?—sat at the center of the "know thyself" project. And we had a complicated algorithm to solve it. First, we gathered promising bands through Radio K, through reviews in *City Pages*, through our older siblings and their friends. Then we located these bands' albums at the record store, and we weighed the names of the albums and songs, the coolness of their album art. Finally, we purchased a CD or two and shared notes on what we heard. This algorithm, though slow and inefficient, gifted us with a deep appreciation for what we listened to. It demanded work, shoe leather and bus fare, time and money spent. Today, when I want new music, I simply tap the Spotify music app on my phone.

Spotify has long pushed personalized recommendations, using everything it knows about what I've listened to in the past to predict what I might want to listen to in the future. In short, it aims to solve the same problem my friends and I faced all those years ago. Quickly, efficiently, Spotify's sophisticated algorithms predict what I ought to like. They're often correct, and Spotify's "Made For You" playlists have proliferated in recent years. The app offers "Discover Weekly," which gives you new songs every week. At the end of the year, it gives you its splashy "Wrapped" playlist, complete with statistics about, and new suggestions based on, your yearly listening habits. It serves up daily sets of old favorites. And, as of 2023, it's presented the AI DJ—a refinement, or maybe the apotheosis, of the app's personalization features. An inhuman disc jockey that speaks to you about the songs it plays for you, the AI DJ fills in the gaps between the songs its algorithm thinks you might like with mindless patter. I still find its uncanny riffs mostly funny and only a little spooky. "Coming up next, a throwback to some of your favorite songs from 2023," it intoned recently. "Not sure if this is a good idea, but here we go." I'm not sure either.

"The beauty of these experiences," Spotify's AI DJ press release reads, "is our ability to deliver the right piece of music for that exact moment in time." There's a bit of hubris here, though you can hardly blame a press release for braggadocio. Spotify thinks it can deliver the right piece of music—thinks it can answer the question that so troubled my high school cohort—with a bit of code. Tacit in this release is who, exactly, was trying to deliver the right piece of music for the right time before the app's algorithms were working: humans.

One of those humans might have been you. One might have been your friend or acquaintance. One might have been a complete stranger, piped in through AM radio. But they all used to offer their own human algorithm to determine what songs and in which order they would appear. Flesh-and-blood disc jockeys at their radio stations, artists with their albums, a roommate at some late-night party, a friend in their car, your own two hands opening a battered case and sliding free a new CD. All these activities share with Spotify the same basic feature: discernment. It's just a question of who—or what—is doing the discerning.

If you focus just on the choice of what to play next, there isn't really a difference in kind between the algorithms dreamed up by Spotify's engineers and those that power human judgment. They both use a kind of "*x* is like *y*" heuristic to decide what follows. Radio station conglomerates and record labels use vast sums of data to inform their "human" algorithms. They conduct surveys and parse through real-time feeds of ratings and listen counts. You can imagine some executive in a glass-walled office somewhere in Los Angeles or New York noting that if people like *this* kind of music, then they might like *that* other kind that shares some features with it. At a house party, your friend might have a good sense of which songs to play, which will drive people into the center of the room to dance. In your car, your partner might suggest a new song they've heard, based on all they know about your past listening.

But the choice of what to play next is perhaps the least important part of tastemaking. When we outsource the hard work of finding the right piece of music for the right time, we diminish the act of listening, turn it from an active process to a passive one. We change it from a deeply social enterprise—one mediated by humans passing knowledge to other humans—into a technological enterprise, one mediated by code. Our tastes, after all, aren't only ours; they emerge from a complex relationship between our listening habits and our milieux, both past and present. While Spotify might do a fantastic job finding music I like, it fails to capture the rich social experience of discovering our tastes and thus ourselves.

Worse, listening to music can be—and has been—commoditized and quantified. This was true even, to a point, before the rise of computer-mediated recommendations; pre-Spotify, studio executives were already doing everything they could to shape their industry with cold, hard numbers. But *quantifiable* listening has only recently reached its zenith, which means our tastes can be, at a new scale, ever more optimized for, our attentions can be ever more hijacked, and our choices can be ever more supplanted. Liz Pelly, in her book *Mood Machine*, adapted at *Harper's*, writes that Spotify has replaced real classical, jazz, and ambient artists from popular playlists with "low-budget stock muzak." This move responds to Spotify's own internal research, which showed that "many users were not coming to the platform to listen to specific artists or albums; they just needed something to serve as a soundtrack for their days, like a study playlist or maybe a dinner soundtrack."

It wasn't always this way. When Spotify launched in 2008, Pelly writes, it "did not start out aiming to shape users' listening behavior. In fact, in the early days, the user's experience on the platform centered on the search bar. Listeners needed to know what they were looking for." But "by the mid-2010s, the service was actively recasting itself as a neutral platform, a data-driven meritocracy that was rewriting the rules of the music business with its playlists and algorithms." Spotify, in other words, went from enabling a user's discernment to replacing it. Instead of a user having to know what she wants, Spotify would simply guess.

In one way, a real issue has driven the switch to computer algorithms from human ones: There's just too much music to listen to these days. Media—recorded music and written words, to say nothing of everything else—have proliferated in the internet age. In 2024, Will Page, Spotify's former chief economist, told *MusicRadar* that "more music is being released today (in a single day) than was released in the calendar year of 1989." In the last few years, according to a UNESCO report, humanity has been publishing over two million books a year. Up until 2010, humanity had published about 130 million books, according to Google research. This means around 20 percent of all books ever published were published in the last fifteen years.

Facebook and its associated apps didn't exist until the mid-aughts, and Twitter didn't exist until the teens, but now we post staggering sums of prose and pictures to their servers. There are, right now, a great many more songs on Spotify, more pictures on Instagram, more tweets on X, more updates on Facebook, and more videos on TikTok than there are seconds left in our lives.

Of course, humans have always created ferociously. But new technologies have enabled more people to make—and store more of what they make—than ever before. This presents a massive sorting problem. We can't possibly watch, listen to, read it, or even know about it all.

To make matters worse, the cultural practices humanity developed to sift through all this media have atrophied. Newspapers cover ever-fewer albums and books even as there are more to engage with. The "heart," "star," and "thumbs up" emojis that speckle our websites are poor compensation. There might be more capturable signal about what to listen to, read, watch, and gaze at now that before, but the tenor of that signal has shifted—away from nuanced, contextualized criticism and into simple, binary reviews. It's the Rotten Tomatoes–ification of media writ large: Why bother reading the criticism when we can just tell you if it's been panned or praised? In other words, humans used to use taste and cultural criticism to sift through the morass, and now we use computers to sift through an even larger morass on our behalf.

This might seem like progress, of a sort. Gatekeepers—largely white men on the coasts who crafted canons of work by, largely, other white men on the coasts—have their problems. As the voices of the marginalized have grown louder, demanding to be heard, and as culture has grown more egalitarian as a result, those gatekeepers' way of sifting through humanity's creative output has seemed woefully inadequate. But it's worth remembering that we've replaced one set of gatekeepers with another: once, mostly white men on the coasts; now, algorithms controlled by mostly white men on the coasts. It's also worth remembering that the hard work of reading the actual criticism rather than the summary of it allows for a much deeper exploration of the media in question.

Digital technologies have allowed us all to access culture at a far greater scale and with less effort than ever before—and that's a good thing. My complaint isn't with the access but with how that access is mediated. AI technologies allow businesses and their profit- and rent-seeking interests to sit between us and this cornucopia in a way they never have before.

When a once-complete medium—the album, say—can suddenly become atomized, that means that medium is ripe for AI disruption. The algorithm of a band deciding on an order of songs on an album does not suit Spotify, which would prefer to craft its own order for you. They do this for two reasons, I suspect. The first is to maximize a kind of facile enjoyment: Spotify knows, or can guess, that its order of songs will keep you listening in a way a band never could. The second is to allow them to inject money-making or money-saving entries into your playlist—an ad, maybe, or a song from a record label they have a deal with. Spotify's user interface encourages you to save a song but not an album or an artist. It nudges you toward the outcome it desires: Spotify mediating your musical taste—capturing it and thus profiting from it.

Spotify isn't alone here, of course. In our social media platforms' first iteration, their users would generally see content only from people they followed, sorted in reverse chronological order, perhaps with an advertisement or two thrown in for good measure. But for those platforms to make money at the scale necessary to support the investments pouring in from venture capitalists, they needed more. More people to create content. More content to show to people. They encouraged us, their users, to house ever more of ourselves on their servers. It worked—though maybe too well. Before long, the crush of content became too much. How could one person keep up with hundreds of friends in the great midteens posting bacchanal? Having given us a functionally infinite amount of content to sift through, the platforms realized they'd need to make sure the stuff that would grab our attention was near the surface if they were going to keep us on their websites and apps.

Soon, they were showing more content from further-afield accounts, unrelated to your friends or even friends of friends, sifting through the detritus of our digital lives with recommender algorithms, which count up how long we spend on different types of content and seek to give us more of the kind that will keep us engaged. Now we have services that are nearly entirely recommendation-based, like TikTok's For You Page or Instagram's Reels or YouTube's Shorts. And of course, like Spotify's AI DJ. These push content they *think* you want—often never consulting you about what you *say* you want. In short, they predict your preferences.

Toil and trouble

For many, the kind of predictive recommendation algorithms popularized online have become almost synonymous with the word *algorithm*, but not all algorithms are predictive. An algorithm, at its core, simply scaffolds a way to solve a problem: It's a blueprint for action. For example, at the *Wall Street Journal*, we once had a list of over one million names of assets and transactions written by about twelve thousand senior executive branch officials in their financial disclosure forms. To describe a sale of a stock, one official might write "Apple Inc.," another might write "Apple," and another might just write "AAPL"—Apple's stock ticker—and on and on. Our problem was that we needed to match all those various descriptions to an actual company. So, my colleagues and I wrote an algorithm that used various matching methods in a particular order to group "Apple Inc.," "Apple," and "AAPL" all together. It took months, trial and error, and what seemed like endless manual validation, but in the end, the database we assembled formed the basis of a Pulitzer Prize–winning investigation that found that one in five senior federal employees were trading stocks in companies that were lobbying them. Algorithms, in other words, can make hard problems tractable.

But though algorithms have many varieties and uses, the one that made *algorithm* a household word is the prediction algorithm. And the trouble with predictions is that they have a nasty habit of coming true. In the first act of Shakespeare's play *Macbeth*, three witches predict Macbeth's rise to become king of Scotland, kindling an ambition that will consume him as he tries to take and hold the crown. Later, the witches warn him to "beware Macduff." So Macbeth, worried, kills Macduff's family—and then is promptly killed by Macduff. These days, more than mere plot devices, prophecies self-fulfill all around us. Algorithmic prediction guides our behavior—what we read, watch, listen to, even who we date—more and more. These companies' algorithms grow ever more complex and opaque to solve a problem they helped create: There's too much stuff to choose from. And what they choose to recommend, more and more, defines what we see.

Sometimes, I feel the anxious pull of Spotify's "Wrapped" playlist. Recently, aware of the limitations of Spotify, I actually bought the composer Will Mason's microtonal chamber album *Hemlocks, Peacocks*, a hybrid of contemporary classical and jazz music. At first, I listened on my computer's built-in music player, but before long, I found myself searching for it on Spotify. Not because it was more convenient, really, but because I wanted credit for my listening. It's like going for a walk without my watch or phone: Why bother exercising if I won't know how many steps I walked, how many calories I burned?

Jeremy Bentham, a utilitarian philosopher from the late eighteenth and early nineteenth centuries, envisioned "the panopticon," a watchtower in the center of a circular prison. Each prisoner's actions might be viewed by the guards—they'd never know if they were being watched—so, presumably, they would always be on their best behavior. "Morals reformed—" Bentham exclaimed, "health preserved—industry invigorated—instruction diffused—public burthens lightened—Economy seated, as it were, upon a rock—the Gordian Knot of the poor-law not cut, but untied—all by a simple idea in Architecture!" Much later, the French postmodernist Michel Foucault popularized the panopticon in *Discipline and Punish*. He suggested it could serve as more than a prison. "Among workers," he wrote, "it makes it possible to note the aptitudes of each worker, compare the time he takes to perform a task, and if they are paid by the day, to calculate their wages." Surveillance both tracks and modifies behavior, in other words—a generalizable, scalable model for the internet age. Not being surveilled by Spotify produced anxiety in me because I have become accustomed to living in a digital panopticon, one with algorithms in the watchtower.

In my time at the *Journal*, I've spent a lot of time helping to report on TikTok's algorithm. TikTok, cleverly, presents only one video at a time, and there are only a handful of actions a user can take on that video. They can swipe away to the next video, they can like it, they can comment on it, or they can share it. For TikTok, these minimal interactions mean that their data is clean, an almost binary signal. How long a user takes before they swipe away for a few dozen videos—this fact alone can put a user into a deep rabbit hole of content. Sometimes that content is benign—my TikTok feed is full of cooking videos and jazz covers. Other times, that content can harm, as with the rabbit hole of pro-anorexia some teen girls fell into, as reported by my colleagues at the *Journal*. The videos we watch—each one of them—leave their mark in the algorithm. Each one pushes us toward more-similar content. TikTok, through their content moderation, would take down some of these pro-anorexia videos, but, it seems, for the teen girls stuck in that rabbit hole, their watch times are remembered even if TikTok takes the videos down. The algorithm keeps a score; the content lives on in its weights.

The right to be forgotten is never more imperiled than in this AI present. You can't "untrain" a model on your embarrassing anime website. You can't make it forget the pro-anorexia TikTok videos, the websites with instructions on how to commit suicide, the bigotry and misogyny of a thousand tweets. You can't even convince Spotify that you don't share your child's fascination with the soundtracks of *Bluey* and *Frozen*. You can, of course, start from scratch, but the AI companies all have their repositories of your work already; getting those behemoths to surface and delete your data requires time, will, and know-how, things many of us lack online.

And this isn't the only way algorithmic surveillance modifies behavior. Critics call the more sophisticated versions of these algorithms, the ones that use machine learning to predict what you might dwell on, *editorial*. The metaphor is drawn from news, where scarce reporting resources are allocated using some editor's inarticulable, but nonetheless deeply felt, sense of what is and is not newsworthy. Recommender systems, in a similar way, direct our scarce attention toward what some machine finds relevant. But where human intuitions about relevance are often humble and vague, the weights machines give to content are absurdly confident and precise. Where a flesh-and-blood editor might suggest you read some story that they think you'll like, an algorithm will predict that you'll dwell on a certain piece of content, say, 97.468 percent of the time. In both human and digital algorithms, the task is to find the next valuable thing. So, it comes down to this: How does one measure value? This question—which has kept generations of editors busy, which is as knotted and ancient as philosophy—our Silicon Valley adepts purport to answer with a clock.

Tick tock

Benedictine monks in the twelfth and thirteenth centuries invented clocks, Neil Postman writes in his 1992 book *Technopoly*, in order to more dutifully praise God. But it didn't take long before the technology "moved outside the walls of the monastery, and brought a new and precise regularity to the life of the workman and the merchant." To Postman, this history shows how capitalism and modernity conquered religion and spirituality. Or, as Meghan O'Gieblyn writes in *Harper's*, "The monastery gave way to the factory. Ritual dissolved into routine." We invented the clock to measure our days, to align our wills to God's. Today, our platforms use clocks to measure us, to align our wills to theirs. Post after post, one after the other, refreshing each time you get close to the bottom, deciding what comes next based on how long you dwelled on the ones before. Fittingly, web developers call this kind of user interface an "infinite scroll," but I prefer the colloquial word for it: *stream*. This word captures the feeling of trying to paddle down one of these infinite scrolls, caught in its algorithm's current.

In 2024, I created a little art piece–as-app called *Leaves of Grasstok*, which presented Walt Whitman's *Leaves of Grass* in a TikTok-style interface. A reader was given one stanza at a time, and I measured how long a reader dwelled on each and determined the next stanza based on only that fact. A quick swipe away would push you into a different part of the poem, but a long read time would anchor you in that kind of material. I was surprised by how well it worked to keep me engaged, even though I made the app to point out the insane ubiquity of these kinds of interfaces.

Our so-called social media giants converged on these kinds of algorithms exactly because they so effectively regulate how we spend our time. But there is nothing all that social about them. I say this knowing full well that a common way of labeling a piece of content has been given the misleading name of "social filtering." If I had used social filtering on *Leaves of Grasstok*, I would have noted the patterns across all the users in what they tended to dwell on, and I would have bucketed each by these patterns: Here are users who like the stanzas about death, here are users who like the stanzas about interconnectedness, and so forth. But social filtering is not social in the way we understand *social* in our day-to-day lives. There is nothing relational about it, only the cold mechanism of clocks synchronizing across abstract buckets of measurable activity.

Though the stream as imagined by our platforms isolates, the same metaphor can also tie the experience of navigating social media back to older ways—much older ways—of understanding time and how we spend it. Ovid, for example, wrote in his *Metamorphoses* that time itself is a river, which "flows on with constant motion." He wrote that "as wave is driven on by wave, and itself pursued, pursues the one before, so the moments of time at once flee and follow." Posts on Facebook, Instagram, X, and TikTok, like Ovid's moments of time, at once flee and follow, "are ever new." They, like time, carry us down a river, wave upon wave driving us forward. In this computer-mediated world, we too flee and follow like the waves. But platforms would prefer we don't think about this human condition—about time's inescapable current drawing us ever closer to our inevitable deaths. The platforms' obfuscations seem almost reasonable when put that way. Who wants to use an app that reminds us that time will always have the last laugh? I suppose I do, given that I chafe when their algorithms enforce a predetermined course down the mighty river of human culture, when they describe a path that hides the world's wonders in all their infinitude.

CHAPTER 8
SEARCH

Deep learning

In 1958, the *New York Times* published an article with the startling headline "Electronic 'Brain' Teaches Itself." The piece touted one Dr. Frank Rosenblatt's invention—a new kind of machine architecture, which he named the "Perceptron." In an early example of tech industry boosterism, the *Times* credulously reprinted Rosenblatt's bluster about his machine's capabilities. "Later Perceptrons, Dr. Rosenblatt said, will be able to recognize people and call out their names. Printed pages, longhand letters and even speech commands are within its reach. Only one more step of development, a difficult step, he said, is needed for the device to hear speech in one language and instantly translate it to speech or writing in another language." Sadly for Rosenblatt, these promises eluded him. But the *Times* wasn't the only institution that bought his hype. The United States Navy, whom Rosenblatt was working for, said it would use the Perceptron to create "'thinking machines' that will be able to read or write." Incredibly, Rosenblatt's machine learning paradigm would, eventually, fulfill his predictions. He just wouldn't live to see it.

Despite Rosenblatt's dying in relative obscurity—considered by most in his field a joke at best or a fraud at worst—his approach undergirds the rise of our current AI systems. "His algorithm is still fundamental to how we're training deep networks today," computer scientist and AI researcher Thorsten Joachims told Melanie Lefkowitz for Cornell University. During his life, Rosenblatt had some problems with scaling his Perceptron beyond a few toy use cases. The issue was, first and foremost, that he didn't have computers capable of training quickly enough. "He was heading on the right track," Joachims said. "He just needed to do it a million times over."

The problem—a lack of computational power—persisted for decades. "At the beginning of the twenty-tens, A.I. was a neglected discipline," Stephen Witt writes at the *New Yorker*. The particular kind of artificial intelligence championed by Rosenblatt, the "neural network," was even less reputable. "Many computer scientists considered neural networks to be discredited," Witt continues.

But that would all change—and shockingly quickly. Over the late aughts and early teens, I played only a few video games. I was never that hardcore about games even before then, but each of the rare times I fired up a new game, it seemed the graphics had gotten more vivid—smoother, sharper, and quicker. In no small part, this was due to the startling improvements in the power and efficiency of graphical processing units—or GPUs, as they're often called. The GPU was long the less-cared-about cousin of the workhorse central processing unit, the CPU, which powers most everything other than graphics. Modern CPUs work very quickly, but they work in sequence, one task after another. They also are generalists, optimized for a wide range of computations—they have to be, since they power so many different kinds of computer applications. GPUs, on the other hand, work in parallel, and they specialize in one kind of math: the same kind of math, by happy coincidence, that powers neural networks like Rosenblatt's.

At its core, the Perceptron was an image classification system. In the showcase that got the *New York Times* all hot and bothered, it was able to say if a square was on the left or right of an image. Fast-forward to the teens, and neural nets were trying to distinguish dogs from cats or sailboats from cars. These neural nets, to this day, all require a vast number of images for training, and, generally speaking, the bigger and better the model, the more images it needs. So where would modern image classifiers get their training data—and how would their performance be measured? Enter ImageNet, which launched in 2009, and which contained "3.2 million cleanly annotated images spread over 5,247 categories," according to its authors. It wasn't long before ImageNet would become the go-to source for training data and for benchmarking. In 2012, "researchers at Google had trained a neural net that identified videos of cats, an effort that required some sixteen thousand C.P.U.s," Witt writes. That same year, in a sea change for machine learning and artificial intelligence, two researchers, Alex Krizhevsky and Ilya Sutskever, bought two GPUs. Those boards, each costing on the order of hundreds of dollars, according to some sources, enabled them to produce "world-class results," Witt writes. They named that neural net—trained in Krizhevsky's parents' basement—AlexNet. It shocked the world. "AlexNet scored so well in the competition that the organizers initially wondered if Krizhevsky had somehow cheated," Witt continues.

And GPUs just kept getting better and better. At the *Wall Street Journal*, my colleague Christopher Mims called the speed of improvement "Huang's Law," after the co-founder of Nvidia, a chip manufacturer specializing in this kind of hardware. "Since 2012 the amount of compute needed to train a neural net to the same performance on ImageNet classification has been decreasing by a factor of two every sixteen months," researchers at OpenAI found in 2020. "Compared to 2012, it now takes 44 times less compute to train a neural network to the level of AlexNet." In other words, the hardware that powers AI has gotten forty-four times better in just eight years, and that's just the hardware.

Today's large language models were born in that happy marriage of our much-improved hardware—and, crucially, our access to much more and much richer data. Rosenblatt's problem wasn't only that he lacked the "compute," as it's called; he also lacked the vast swaths of data the compute would be used for. Not just a few thousand pictures, but millions of them. But couple the wide availability of ever more powerful GPUs with the cornucopia of words and images and videos—bytes upon bytes of the stuff—that we now host online, and you have the exact ingredients Rosenblatt was missing for neural networks to finally make good on their promise.

That second ingredient—data—has, in many ways, proved more problematic than the first. It has required the proliferation of a particular kind of program, a group of bots called crawlers. At some level, these crawlers are beneficial; not unlike bugs in the garden, their presence marks a healthy ecosystem. Google and other search engines, for example, send these crawlers out to see what the web looks like. Search engines use crawlers to "index" the web—to read every page online, following each link in turn, and store information about what those pages contain. Without them, Google couldn't answer your queries. In my job as a reporter, I've written more crawlers than I can count. I've let them loose on government web pages to see what new administrations have changed. I've sent them to vacuum up important press releases. I've allowed them to gorge themselves on biographies of corporate executives. Crawlers power various web security plays, and the Internet Archive uses them to save vast swaths of the internet. "And as of late, they serve one more purpose," Shayne Longpre writes at the *MIT Technology Review.* "Companies such as OpenAI use web-crawled data to train their artificial intelligence systems, like ChatGPT." As of early 2025, Longpre notes, crawlers now account for "half of all internet traffic." If trends continue, it won't be long before they, not humans, account for most of the internet's requests.

Creepy crawlers

To see why this matters, it's important to know how the internet actually works. Generally, some "client," a machine—your laptop or your phone, say—requests some code that describes a website. Another machine, called a "server," responds by sending back the code you requested. Finally, your browser, sitting on the client, interprets (or "runs") the code, actually building the website from the blueprints sent by the server. Modern websites often pastiche code together from multiple different sources, which means your one request might be fulfilled by several servers all working together. Each of these fulfilled requests uses various resources—servers use "compute" when the server itself needs to run some code to respond, and they use "bandwidth" when they actually send the data representing the HTML or JavaScript code (or image or video files). The more compute and bandwidth used, the more expensive the request will be to fulfill. Common requests are often "cached"; there are even entire services that precompute and cache these requests at data centers all around the world. The company Cloudflare, for example, has made a name for itself—and oodles of cash—by operating a vast network of servers that cache responses to ordinary requests. Cloudflare, according to Longpre, supports about a fifth of all web traffic. All this is to say, crawlers, humans, and the sites both groups touch used to operate in a kind of equilibrium.

When I write crawlers, I'm generally circumspect about their use. Each request a server needs to fulfill, particularly on pages that haven't been cached, costs money, power, and compute. So, I have my crawlers make requests consecutively rather than concurrently; instead of hammering some poor government server with a few dozen requests all at once, I spread my requests out over a couple of minutes. When I'm requesting little-used pages, I'm often even more careful, even slower. But the crawlers used by data aggregators aren't always so kind. And while humans might request things that humans actually want to read, bots request, well, everything. On Wikipedia, which was built for humans, its servers cache the code for often-requested entries, making those requests relatively cheap to respond to. But now, "a significant amount of our time and resources go into responding to non-human traffic," Wikimedia Foundation staffers Birgit Mueller, Chris Danis, and Giuseppe Lavagetto write. And, they say in that typically understated Wikipedia way, "while human readers tend to focus on specific—often similar—topics, crawler bots tend to 'bulk read' larger numbers of pages and visit also the less popular pages. This means these types of requests are more likely to get forwarded to the core datacenter, which makes it much more expensive in terms of consumption of our resources." In other words, the crawlers have gotten out of control.

As large language models grow ever more voracious—hungry as they are for human-produced words, images, and videos—many websites have thrown up new and more aggressive defenses against crawlers. A whole cottage industry has sprung up to build anti-bot protections, and these fortresses, while a logical extension of the crawler–server arms race, threaten to close off the once-open pastures of the web. "Small creators, such as visual artists, YouTube educators, or bloggers, may feel they have only two options: hide their content behind logins and paywalls, or take it offline entirely," Longpre writes. "For real users, this is making it harder to access news articles, see content from their favorite creators, and navigate the web without hitting logins, subscription demands, and captchas each step of the way."

As with any ecosystem, the problem is balance. When a real person needs to find and access some information online, some number of crawlers is necessary. How else can a search engine know what's at some URL? But, as operators close their sites to the ever-growing swarm of bots, they are not just closing sites off from search engines and web preservationists; they're making it harder for regular people to access content that used to be readily available. A careful gardener can use insects to add fertility to their plot of land, but if one kind of bug takes over, it can spell doom for their crops.

Neural networks like the Perceptron eventually produced the large language model, those stunning pieces of technology that can output language. And as the great cathedrals of neural network–style architecture, compared to AlexNet's lowly chapel, large language models require even more compute, and even more data. "Companies racing to develop more powerful artificial intelligence are rapidly nearing a new problem," my former colleague Deepa Seetharaman wrote in 2024 at the *Wall Street Journal*. "The internet might be too small for their plans." Large language models "require larger oceans of information to learn from," Seetharaman wrote. "That demand is straining the available pool of quality public data online."

The crisis might have been averted—or at least delayed. Large language models, researchers discovered, are able to produce and train on what's called "synthetic" data—when one AI system produces language that is fed back to another AI system. It turns out that you still need to temper that synthetic data with human-generated data, otherwise it can cause the model to "collapse"—to progressively degrade with each training session until the models "become useless," researchers Matthias Gerstgrasser et al. write in a 2024 paper. They show that when you replace the real data with synthetic data, it can cause the model to collapse, but when you accumulate real and synthetic together, you can spare your model. How convenient, then, that so much of our digital lives have been kept online, ready to be "accumulated" into AI systems.

So, the need for human-generated language has diminished but not disappeared. And as companies continue to race to find new caches of words to stave off model collapse, they are also racing each other. Competition might breed invention, but not all invention is, strictly speaking, above board. Meta's Llama 3 large language model, meant to rival OpenAI's ChatGPT, Google's Gemini, and Anthropic's Claude, "would need to be trained on a huge amount of high-quality writing to be competitive," Alex Reisner writes at the *Atlantic*. "Acquiring all of that text legally could take time. Should they just pirate it instead?" Spoiler alert: They did. "Meta employees turned their attention to Library Genesis, or LibGen, one of the largest of the pirated libraries that circulate online," Reisner continues. LibGen contains tens of millions of research papers and millions of books—including, as I discovered to my discomfort, my first book, *Lessons and Carols*.

"Eventually, the team at Meta got permission from 'MZ'—an apparent reference to Meta CEO Mark Zuckerberg—to download and use the data set," Reisner writes. AI companies argue that their use of crawlers to pull down, basically, the entire internet constitutes fair use, that the normal rules of copyright don't apply. But it's hard to defend pulling books from an explicitly pirated source, especially given Meta's exploration of licensing and paying for the books from publishers. It's not that they didn't know better, in other words; they just wanted the data more quickly and cheaply than publishers were willing to offer it.

Model collapse isn't the only way AI models are eating their own tails. They're also undermining the institutions that enable human writing in the first place; one of the basic utilities of the web, one that enables what we think of as the web to exist, is under threat. I'm talking about search. "If you've tried to search for a recipe or product review recently, I don't need to tell you that Google search results have gone to shit," Dmitri Brereton writes in a provocation at his Substack. As someone who often makes these kinds of searches, I'm hard-pressed to disagree; the quality of Google's search results has gotten worse. And it's not only product reviews. Much of my job involves scouring the little-traveled corners of the web for sites and documents on obscure topics—for older pages, or pages off the beaten track. Google Search, in my estimation, still produces sensible results when I want articles from major news sites or well-trod concepts. But at the intersection of obscure and problem-focused—that's where it falls down. "The first few non-ad results are SEO optimized sites filled with affiliate links and ads," Brereton continues, tracking my experience perfectly.

It is the very model of a modern AI company

Tom Scocca, at *Flaming Hydra*, writes that he wanted to look up Mikhail Baryshnikov's Latvian origins. On his phone, after making his way to the search results page for "Mikhail Baryshnikov," he found "a blob of unreadably generic text" with a sentence-long overview biography, some boxes with recent links about Baryshnikov, and a box titled "quick facts," which noted the ballet dancer's Latvian heritage, which did not cite anything. "By now I've learned to usually scroll past everything at the beginning of a Google search result, but I was on the phone and in a hurry, so instead I just spotted the word 'Wikipedia' at the bottom of the text-blob and tapped on that," Scocca continues. Clicking on that revealed a text pop-up with AI-created language summarizing the Wikipedia entry, "displaying the sort of link to the Wikipedia page that, a few years ago, would have been the immediate result of searching 'baryshnikov' on Google." Finally, Scocca clicked on the link embedded in the machine-written summary, and he was taken to the Wikipedia entry—except, "Google had chosen to send my browser to a passage lower down on the page," where the search engine had taken it upon itself to highlight some text that it felt was important. "Four clicks in," Scocca writes, "rather than directing me to information people had published about Baryshnikov, Google was still directing me to what its own machinery had decided must be the answer that someone Googling 'baryshnikov' would want." In other words, Google has replaced its agency for ours.

It used to be that googling was a bit of an art. You had to know things like exact terms should fall within quotation marks, and you had to learn and think through the correct Boolean operators—*and*, *or*, and *not*. But Google, years ago, began moving toward trying to suss out a searcher's intent. Now, "Google is trying to scan the query, make meaning from it, and surface pages that it thinks match that meaning," Charlie Warzel writes at the *Atlantic*. "Despite being a bit sci-fi and creepy, the shift might feel like a loss of agency for searchers. Search used to feel like a tool that you controlled, but Google may start to behave more like, well, a person—a concierge that has its own ideas and processes." Warzel is describing Scocca's experience in different words: the breathless way Google's Search offers up its own interpretation of the underlying internet, like a cloying intern. This mode of interaction presents two kinds of problems for the experienced searcher. First, a machine's assumptions about what exact question you want an answer to can be wrong, and second, even when it's correct about what you want, the labor Google's algorithms are saving might be the most necessary labor a researcher can do—discerning what is and what is not relevant.

There are cases when I don't really want to use Boolean operators and consider deeply what I'm looking for. But in its rush to provide that experience, Google has taken away even the choice of a more sophisticated search. It has even gone so far as to replace actual searching with generative artificial intelligence, with a large language model, and as Google encourages us to read their summaries rather than click through to the text from which the summary was drawn, they are threatening the livelihoods of the creators of source texts who can't make money off crawlers in the way they make money off real humans. In other words, websites' actual human traffic is being throttled even as crawlers choke them. The irony is plain: Large language models and their predecessors—neural networks—were enabled by the internet. Now, they're destroying it. We humans—all our labor, blog posts, forum discussions, news articles, books, papers, all of it—fed the gaping maw of AI, and now it's biting the hands that feed it.

That's the "supply side" problem with the AI-ification of search, but there's trouble for its users too. To steal a metaphor from friend and fellow technologist Nick Hall, when we research, we are driving a car from point A to point B. But when we send a model to create a summary, to rip it from its context and deliver it to our eyes, we are not getting from point A to point B faster;

instead, we've sent a drone to go to point B for us and take a nice picture we can look at. In this way, when Google summarizes for us, our minds don't go on the journey. Simply put, they don't learn. I should say, we've long relied on summarization to learn. Much of my job requires hours and hours of research. I would be a terrible researcher if I didn't know when to skim and when to read, when to check out a summary and when to buy the whole book. So, the trouble isn't the technology of machine summarization—not exactly. The trouble is the assumption that machine summarization is what we want all the time, the way Google and others put it on top of every search, after every keystroke, in answer to every query.

The author and educator Phil Christman calls AI writing "the dead algorithmic repetition of what already exists." This rings true to me, though it's not to say that dead algorithmic repetition of what already exists means useless or new. When scanning thousands of documents for one concept or idea, machine-generated summaries can help make the haystack that needle is in a bit smaller; it can help order my reading. And, of course, as Christman insists, this kind of writing existed long before AI as "the thing your kids were already doing on those damned tests, but more effortfully before the machine existed."

Using AI to read and write (which is to say, to think) so that you don't have to might be the apotheosis of an instrumentalized way of understanding those great human activities, but instrumentalized writing isn't new. In education, reading and writing are often "reduced to smaller and smaller mechanized gestures and procedures that teachers and students are forced to exhaustively and exhaustingly repeat by an external test-imposed curriculum, all at the expense of actual book reading," Christman writes.

I was a thoroughly mediocre student, but the classes I loved, I really did love. I remember Mr. Bazzett's steady voice guiding us through Louise Glück's collection of poems *Meadowlands*, and Robert Hass's collection *Praise*. I remember Mr. Hollister letting me put the Velvet Underground's album *White Light/White Heat* in conversation with F. Scott Fitzgerald's unfinished masterpiece, *The Last Tycoon*. I remember Ms. Landis teasing out a line of music in Shakespeare's *Hamlet* so that I could discover more of its rhythms and harmonies for myself. Not one of these experiences could have been aided by more process, be it AI or human. The beauty of these experiences rests precisely in their outsideness—outside process, outside structures, outside myself.

When AI is used to do our thinking for us, it encases our thoughts in its shrinking walls, and it becomes harder to see our way out. "My current problem, what I struggle with," Karl Ove Knausgård writes in *Harper's*, "is that I cannot find an outside to technology. It is as if the outside had disappeared, as if it were no longer a possible place." That sense, that an "outside" is no longer even possible, comes when our agency is denied. But, like Hamlet's friend Horatio, our AI companies—Google and all the rest—put us and our digital lives under glass, as though we were one of those little birds ready to be eaten at a fancy dinner party. Still, as Hamlet says: "There are more things in heaven and earth, Horatio, / Than are dreamt of in your philosophy." There are, but the AI giants seem determined to enclose our dreams too.

"It's the new millennium," the poet and essayist Susan Howe writes in *The Quarry*. "Post 9/11, spangled bleeding banners, war's carnage, the global War on Terror, Guantánamo, metadata relationships, fracking, plastic bags, nuclear power plants, climate change, global warming, black holes, possible human extinction." Howe's litany, written in 2015, couldn't include large language models, though maybe it should. As our creations continue to fulfill Frank Rosenblatt's prophecies, seventy years removed, the dawning unreality saturates, coloring too vividly, too brilliantly. "If we have nothing but truth to leave," Howe continues, "how do we distinguish ideas of what we were from ideas of what we are in a vibrant contemporary compost jargon trash landfill"? We are, indeed, in the trash—in the collected detritus of decades online. The question of what we are, especially given the potency of generative artificial intelligence, demands an answer. Perhaps we are, as Howe's passage suggests, Oscar the Grouch living inside a landfill. The AI models certainly treat us that way; their reliance on our ephemera means they must.

The cheery violins and merry trumpets in the opening of Bach's Brandenburg Concerto No. 2 bounce off the beat, racing forward. These instruments appear as the first music on the golden records sent inside the Voyager spacecraft—humanity's best time capsule—and they give way to a sequence, circling by perfect intervals in the bass while the melodic voices counterpoint above. Later, the trumpets will rise in minor over a piano in Louis Armstrong's "Melancholy Blues." The court gamelan of Java rings beneath singing voices, drums beat, a flute whispers. Chuck Berry hammers out the iconic guitar riff of "Johnny B. Goode." Berry's inclusion proved a bit controversial. One folklorist on the Voyager project called the music adolescent, according to Megan Gambino at *Smithsonian* magazine. "There are a lot of adolescents on the planet," Carl Sagan is said to have responded.

When I imagine some future alien society coming to Earth and finding a working copy of a large language model trained on all the wealth of human experience—and all the garbage too, not just adolescent but utterly juvenile—sadness takes me. We have no Sagan to oversee the choice of how we represent humanity, only the rapacious interests of "MZ" and his ilk, signing off on rank piracy. We have only the cloying summaries of a dancer's life. We have only a cold, sterile inside.

CHAPTER 9

BELIEF

On a reporting trip for the *Wall Street Journal*, I spent almost a week trekking from Portland, Oregon, to Sandpoint, Idaho—a tourist trap of a town on the north side of Lake Pend Oreille. A colleague and I, plus a pair of diving researchers, drove down the Columbia River from Portland to Lewiston, Idaho, then all the way up to the Pend Oreille River and along it to the lake and Sandpoint. Everywhere we went, we stopped where we thought old lead-sheathed telecommunications cables dipped into the rushing expanse—cables we suspected were leaching toxic lead into the water and soil around them. At every crossing, we boarded a little boat (I wanted to name her *Rupert's Folly*) to look for signs of the cables on the shore.

Before we had left, my fellow reporters had made Freedom of Information Act requests from every US Army Corps of Engineers office in the country for the original permits—sometimes a century old—that gave the various telecommunications companies the right to run their cabling beneath navigable waters across the country. My job was to process the tens of thousands of pages of antique documents, to clean up the often fuzzy scans, to make their text searchable, and to use a combination of machine learning and good old-fashioned data entry to parse out their exact locations. By the time we flew to Portland, I had stitched together a massive database, complete with the old maps from the original permits and all their annotations—and I had thrown it all into a little website for the reporters on the project that showed where everything was.

While *Rupert's Folly* sliced through the choppy water, the wind at my face, I'd break out my laptop and a Wi-Fi hotspot, connect to the sparse internet, pull up the permits and maps, and look for landmarks. I rarely won the little game the four of us played: The first person to identify a cable in the thickets of dry shrub and fallen tree limbs on the side of the river would get bragging rights. By the time we made it to Sandpoint, I was exhausted from the long days in the sun and on the water, from the joyous unreality of being paid to ride down the breathtaking river when, normally, I'm paid to sit at a computer.

Missing my partner and my daughter, I left the night before the others rather than spend another day on the road. I took a cab back so I could hop on a red-eye at the little airport in Spokane, Washington—the nearest airport that serves Sandpoint. By the time the cabbie pulled up, the last dregs of sunlight pouring out thickly across the lake, I was dead on my feet. The cabbie told me it'd be a long ride, and he asked me if I'd like to sit in the front seat. I said I didn't mind. Lulled by the sight of the rippling lake as we crossed its Long Bridge, I almost didn't notice that the cabbie was showing me the handgun he had holstered at his hip.

"Just so you know you're safe," he said.

I wasn't sure how his gun would keep me safe. Were there marauders on the road to Spokane? Still, I thanked him for telling me. It was an inauspicious start to the two-hour drive.

He began by quizzing me about my thoughts on guns. Had I ever used one? I told him I'd never even held one. What did I think about the politicians, the Democrats, who wanted to ban them? I said I wasn't sure any Democrats with actual political power were really trying to ban them. Did any of my friends own guns? I told him the only friend I had with a gun also supported Senator Bernie Sanders for president. Was I pulling his leg? No, I wasn't. He asked me what I did for work, and I told him I was a journalist.

"The world is spinning off its axis," he told me. "The media won't tell you that." He meant it literally. Scientists, he insisted, had discovered that the world was turning faster, irregularly. "Look it up," he said.

The sun had fallen behind the horizon by then, draining the world of color. Maybe the world really had spun off its axis, the strange landscape, so different from anything I'd ever seen, seemed to whisper. The constant challenges from the cabbie, who was jumping from topic to topic before I could get my bearings, only added to my disorientation: guns, cops, the COVID-19 vaccine, which would leave me infertile, he declared, my body broken.

Eventually, I decided our conversation about guns would be more solid ground than this litany of conspiracies. So, I asked him why he wanted his guns, what he got from them. This wasn't only a protective move; I was genuinely curious. He told me he had been robbed when he lived in Spokane, while he was home, and he wanted a way to protect himself. He described the break-in in exacting detail, like he couldn't forget the scene.

Driving with his knees, his eyes never leaving the road, he extracted an iPad from the bag next to him and mounted it on the dashboard. From YouTube, he pulled up a video about the importance of the Second Amendment, the necessity of protecting yourself. "Be a defender," the narrator encouraged. After, the cabbie asked me what I thought. I told him, honestly, I wasn't sure what I was supposed to be defending myself from. He looked incredulous, and I wondered about the differences in our lives. If I had been robbed, and if I had stumbled onto these videos that could teach me how to never be afraid again, would I be nodding in agreement rather than bemusement? Would I, too, have come to believe that the COVID-19 vaccine, which I had taken with relief, would break me?

Online, every theory can be supported. Every notion carries the imprimatur of authenticity, the ring of reality. Lies live on YouTube, right next to videos by real scientists. Lies are served in Google search results, right next to authentic sources. Lies are laid out on social media in the same slick fonts as the truth.

Since that cab ride, the world has spun on, ever more punch-drunk. If anything, it's spinning faster. On Facebook, bots post AI-generated images of Jesus, his body made up entirely of shrimp. Shrimp Jesus prays before a congregation of other bots who respond with prayer emojis. I can't look away, and Facebook's algorithm begins to serve ever more AI slop, like I'm a pig being made ready for slaughter.

On TikTok, videos voiced by human-sounding robots assure me that President Trump is having a love child with a Republican congresswoman, that Barron Trump keeps crying, embarrassed that his father wears feces-filled diapers. These fake voices read from scripts bearing all the hallmarks of a large language model. I watch them again and again, for work this time. Every time TikTok's spam filters take one bot down, the spammers spin another up.

On my phone, I receive text after text from scammers who tell me that I've left bags in their BMWs or that I've missed our lunch date or that a trip to the beach would be nice. On X, I am messaged by young men (bots) who want me to buy cryptocurrency from them. On Instagram, I am messaged by young women (bots) who want me to date them. The web bleeds together into a kind of unreal, inhuman montage, the internet's promise of human connecting to human turning sour in my mouth, we're choking on it.

In October 2023, I sat on a plush couch in my home in New York's Hudson Valley, the leaves on the trees shocking me with their merry orange and bloody crimson, and I read the first news articles about Hamas fighters' brutal incursion into Israel, the war crimes they committed, the children and families they murdered. Soon, I knew, the Israeli military would begin a retaliatory campaign in Gaza, where residents, cut off from the rest of the world by Israel, were already living in desperate conditions. In the months that followed, I swiped through video after video of dead children in Gaza, their broken bodies clutched by their wailing parents, bombed by weapons my government supplied. A doctor in Gaza described the razing of his hospital. A young woman looked into the camera and told me about moving from supposed refuge to supposed refuge, followed by our bombs. I saw young IDF soldiers take pictures in front of blown-out homes.

War, its horror, is an old story. The closeness of it jars—the presence of it in my comparatively decadent life. But what shook me most was this: Politicians of all stripes assure us that to agitate for a ceasefire, for an end to the flow of guns and bombs and money from our government to Israel's—to make these videos stop, in other words—is foolhardy at best and cosigning Hamas's brutality at worst. They say that to call for an end to the carnage is itself a kind of a deep violence, that it demonstrates antipathy toward the Jewish people. This logic confuses me, even as it comes from a bipartisan elite. American social life has failed to grapple with our role in the decimation of Gaza. American social life has failed to grapple with the world as it is—spinning off its axis.

Even paying attention to the unfathomable pain in Gaza is suspect. "How did we reach a point where a majority of young Americans hold such a morally bankrupt view of the world?" Republican Representative Mike Gallagher, who advocated for banning TikTok, asks at *The Free Press*. "Where were they getting the raw news to inform this upside-down world view?" His answer: "social media and predominantly TikTok." It's true that social media can disorient our moral lives; how could it not when we hear the dying scream out from our phones' tinny speakers? How could it not when our leaders, like Gallagher, say that feeling outrage at the deaths of so many is really "celebrating Hamas's murderous cause"? The people of Gaza are dying deaths from weapons taxpayers like me fund, deaths that I would say are unimaginable, except that I can see them with my own eyes from my living room.

The internet invites the cruelty of the world in, through the threshold of my house, up onto the couch next to me. In this way, the medium of the web can inspire moral outrage. It can also inspire action. Digital technologies can organize mass protests and mobilize political pressure and encourage real action. This is true, even as it is also true that the medium of the web is wearing down the fabric of civic life and patching its tattered places with Instagram meme pages, strange Reddit forums, and ill-moderated Facebook groups. The web is mostly dark corners, it sometimes seems—just-for-you algorithms and random, stumbled-upon websites. In these places, the conversations push themselves to the extremes, their norms designed to radicalize, their interfaces designed to give the illusion—or maybe unmask the reality—that everyone, everywhere, holds some kind of fringe belief, that we all have some gnostic truth, forbidden by the powers that be.

By February 2020, a novel coronavirus had made its way to the United States, sickening several. Panic soon followed. Wipes and hand sanitizer went missing from the shelves of pharmacies across the country, and you couldn't find a face mask to save your life. The then-surgeon general Dr. Jerome Adams took to Twitter. "Seriously people—STOP BUYING MASKS!" he wrote in a since-deleted post. "They are NOT effective in preventing general public from catching #Coronavirus."

In March, just as the World Health Organization declared COVID-19 a global pandemic, the CDC was agreeing with Dr. Adams. "CDC does not recommend that people who are well wear a facemask to protect themselves from respiratory illnesses, including COVID-19," their now-inaccessible FAQ read. In those early days of COVID, public health officials were speaking with one voice: Masks were ineffective for the general public, and if they continued to buy them, the shortage would prevent those who needed them from getting them. "If health care providers can't get them to care for sick patients," Dr. Adams's tweet continued, "it puts them and our communities at risk!"

This guidance wasn't to last. It wasn't long before public health officials had reversed course. "CDC recommends all people 2 years of age and older wear a cloth face covering in public settings," a new

CDC website read. As COVID began its spread, we didn't know what we didn't know, which included quite a bit, it turned out—but public health officials still didn't turn off their metaphorical caps lock.

COVID wasn't the only thing spreading in 2020; falsehoods, too, were in the air. Social media platforms began throttling "harmful" misinformation about the virus. Twitter declared that it was broadening their "definition of harm to address content that goes directly against guidance from authoritative sources of global and local public health information." Facebook, during public health emergencies, "in partnership with global and local health authorities," would be removing "false claims which discourage good health practices related to COVID-19"—which included false claims about the utility of wearing a face mask. Given the changing guidance from those authoritative sources of health information, you'd be forgiven for wondering just what constituted a "false claim" on social media.

To be charitable to our public health officials, they didn't know the full extent of COVID's capabilities, and there was a real shortage of face masks in those early days. It's easy for me to second-guess their strident but shifting asks, to try to play armchair Dr. Fauci. But no matter how you slice it, the *feeling* of confusion was real, and it almost certainly contributed to the rising conspiratorialism that followed everywhere the virus traveled. Did Bill Gates cause the pandemic through the spread of 5G cellular phone towers? Did the US military seed the virus inside China to blacken their government's reputation? Would the vaccine really leave me infertile, as my cab driver said?

Conspiracy theories "get the facts wrong but often get the feelings right," Naomi Klein writes in her book *Doppelgänger*. The facts are plain: We should have worn masks, and neither billionaires nor phone services nor the US military wanted to spread the illness. But the feelings are also plain: COVID-19 was confusing and baffling, destabilizing to both our bodies and our social structures—and the strident waffling from public health officials only added to the sense of dis-ease.

I used to be a happy connoisseur of flat earth memes, a conspiracy theory that carries some startling implications. The moon landing was faked, for example. Airlines and shipping companies are lying to us all the time. There's an ice wall, guarded by NASA, that wraps around the flat plane on which we make our homes. And so on. It's hard for me to take it too seriously, especially since the memes are so juvenile and almost wholesome in their silliness. "Water can't hug a ball," one reads. "Round earth is a lie." I should have known better, though. Even the flat earth theory ends up rooting itself in perhaps the oldest conspiracy theory: anti-Semitism, the hatred and fear of Jewish people. "Globe . . . Globalism . . . Any questions?" one meme asks. "The globe is a lie built on Satanic Freemason Kabbalistic Paganism," reads another. When you wonder who is lying to you, the false, dangerous answer many reach for—as it has been for centuries—is "Jewish people."

In October 2018, a man attacked the Tree of Life synagogue in Pittsburgh, Pennsylvania, killing eleven, shouting that "all Jews must die." The attacker had marinated in a stew of anti-Semitic conspiracy theories online long before he struck. He believed, as many on the white-supremacist right do, that Jewish people support mass immigration of nonwhite people to the United States and Europe in order to commit a kind of "white genocide." (Just writing that sentence made me feel as though my brain were rotting.) The attacker found his voice on Gab, a Twitter-like social media platform popular with the "alt-right"—a phrase that attempts to rebrand white nationalism.

This is the two-faced internet. We see more of the truth than ever before. Footage from body cameras shows the world the police harassing and even killing Black Americans. Amateur cell phone videographers post unfiltered videos of social uprisings in Ferguson, in Hong Kong, in Tunisia. The people of Gaza live stream their pleas for help. But we see more untruths too. The world is not flat. George Soros is not puppet-mastering demands for a "woke" society. "The Jews" do not control the media.

These dueling phenomena are related. When our political elites tell us that what we observe is not what we're seeing, that what our moral compasses tell us is untrue, they set the stage for all manner of conspiracy theories. Politicians say an economy ravaged by automation and laissez-faire trade policies is good, actually, that the neoliberal, technocratic order has made the pie bigger for everyone—a rising tide, after all, raises all boats. It might be true that our GDP grows, but what do we say to the hollowed-out towns of Middle America? When elite institutions ignore their plight, they fail to prepare for the rising fear of immigrants, which, through the sickening filter of white-supremacist forums, becomes the fear and hatred of Jewish people. As young Americans learn about Israel's actions—backed with American dollars and weapons—in Gaza, political discourse about our unwavering support becomes ever more off-limits. Is it any surprise that they are outraged?

In his book *Believe*, Ross Douthat uses the term "Official Knowledge," which means, roughly, information about the world around us that elite institutions cosign—information, as he says cheekily on *The Ezra Klein Show*, "considered normal and respectable in publications like the *New York Times* and Ivy League universities." For Douthat, Official Knowledge means an assumption of disenchantment or a kind of default "practical atheism," as he writes in *Believe*. "Disenchantment in this specific sense is a modern consensus, an intellectual default, even for the sort of academics and journalists and Wikipedia editors who might show up for church on Sunday . . . or otherwise personally defect from disenchanted premises."

In the book of Acts, Saul was persecuting the early followers of Christ. While on the hunt, traveling the road to Damascus, he fell to the ground as a great light from heaven flashed around him, blinding him. For three days after, he neither ate nor drank, and he was led by the hand into the city where a disciple healed him. "Immediately, something like scales fell from Saul's eyes, and he could see again" (9:18 NIV). Saul would change his name and become Saint Paul, to whom theologians attribute about half the books of the New Testament. This is the anatomy of a reenchantment: The fresher the conversion, often the more zealous the belief, which is no surprise. When you think you are first seeing the world not as *they* would have you see it but as it really is: a blinding, holy beauty.

I'm using Douthat's term a little more broadly than he intended—to speak to the way elites police beliefs both secular and religious. But the casting aside of a dimmed world in favor of bright revelation needn't be religious, strictly speaking, nor does reenchantment require a deity. Indeed, what we see with Saul on the road to Damascus, we see with new adherents to any kind of marginal worldview. The scales fall, and the world resaturates, reenchants in the warmth of a sun unshaded by Official Knowledge. Officially, we ought not to have worn masks, until we ought to have worn them. Officially, neoliberalism created a strong economy. Officially, what's happening in Gaza is necessary for the self-defense of the Jewish people. But the internet easily challenges all these official proclamations—for good and for ill. And as the minders of official knowledge demand we all ignore things we can plainly see, more and more of us will reject not only the official knowledge but also the systems that produced it, igniting a jubilee of mistrust of elites across science, economics, and politics.

Conspiratorialism has long ailed the American body politic. In his 1964 essay in *Harper's* on "the paranoid style of American politics," Richard Hofstadter wrote about the "heated exaggeration, suspiciousness, and conspiratorial fantasy" besetting our nation since its founding. In the late eighteenth century, Americans were fighting the Bavarian Illuminati. A few decades later, the Freemasons. Not much later, it was the Catholics. And of course, through it all, there was the struggle against scheming international bankers—who give all this theorizing a Jewish-coded edge.

Beliefs about these imagined conspiracies root themselves in the fertile ground of dispossession, Hofstadter writes, in the idea that "old American virtues have already been eaten away by cosmopolitans and intellectuals"—a complaint that still has purchase today. Hofstadter notes that mass media proved a boon to these theorists. In that age, "the villains of the modern right are much more vivid than those of their paranoid predecessors, much better known to the public," he wrote. In the age of digital media, the villains of American conspiracy theories aren't just vivid; they are practically embodied on our phones and screens. So no, the internet alone hasn't caused our paranoid political environment; it merely makes it easier to pull the conspiracies from the rich vein of paranoia that runs through American political life. And as inequality—geographic, economic, educational, etc.—rises, peeling elite institutions away from the rest of the country, that vein only seems richer.

But the medium of the web still matters. Misinformation moves quickly online, generating engagement wherever it spreads. A paper by MIT researchers, published in *Science* in 2018, found that Jonathan Swift was correct: "Falsehood flies, and the Truth comes limping after it." According to Soroush Vosoughi and his MIT coauthors, who studied millions of tweets, "Falsehood diffused significantly farther, faster, deeper, and more broadly than the truth in all categories of information." Novelty, Vosoughi et al. found, drove much of falsehood's spread; a surprising nugget of untruth traversed Twitter in ways boring old truth couldn't. But the intersection of something new, on the one hand, and something fear- and disgust-inspiring, on the other, proved even more potent.

In the fever of the early aughts, the one that gripped the American body politic in the wake of the September 11 attacks, Official Knowledge had it that Iraq harbored weapons of mass destruction. The organs of American political life credulously repeated the claims made by the Bush administration, demanding fealty to that lie. The budding antiwar movement, then, had to find its purchase in the rocky soil on the outskirts of mainstream discourse. Every march I attended, every vigil and protest, included a loud contingent of so-called 9/11 truthers: people who believed the attacks on the World Trade Center and the Pentagon were known about in advance and allowed by President Bush—or even actively planned by him and his administration. The planes that smashed into the buildings weren't the only peril faced that day, they argued; jet fuel can't melt steel beams, the line went. And, of course, those who adhered to this belief also often soaked it in the acid waters of anti-Semitism. Jewish workers in the Twin Towers called in sick en masse, some said. As a teen, I found the "truthers'" presence in the antiwar movement mostly confusing and gauche, kind of embarrassing. Now I see their poisonous lies as dangerous—and tragic. The wrongness of their facts was matched only by the correctness of their feeling: The Bush administration and their media cheerleaders were not to be trusted. They were being lied to, just not in the way people thought. How much energy was spent on a nonexistent, all-powerful, secret cabal when, in reality, the conspiracy fell to a clique of bumbling ex–cold warriors, drunk on their own power, and a stampede of journalists, easily spooked into groupthink?

It's easy to forget the stakes of the panicked patriotism that followed the attacks of September 11. It wasn't only the American flag being raised from building after building, the pins decorating the lapels of politicians, and the renaming of *french fries* to *freedom fries* in the US House of Representatives' cafeterias. It was also the enforced silence of critics of US foreign policy and the dramatic rise in anti-Muslim hate crimes. In a 2011 paper, Ilir Disha and his coauthors note that there were twenty-eight "hate incidents" reported by the FBI in 2000. In the six months following the attacks, there were 117 reports of hate crimes against Muslims in New York City alone. And of course, the stakes didn't stop at the border. We must count the US service members and the hundreds of thousands of Iraqis killed in a war launched under false pretenses.

As the nation rallied around the flag, I endeared myself to my classmates with a recitation of Ani DiFranco's spoken word piece "Self Evident" at an open mic night in my high school's art classroom. While some friends played sparse jazz behind me, I repeated DiFranco's elegy for an America skeptical of imperial power and her hopes for a multiracial, multiethnic democracy that embodies an egalitarian refutation of hate. But it wasn't only that. "And we hold these truths to be self evident: / Number one, George W. Bush is not president," I intoned, refusing the election's legitimacy. "The media is not fooling me." I was no 9/11 truther, but I was testing a divorce from the institutions of American life that produce Official Knowledge.

I sometimes wonder what the run-up to America's war on Iraq would have been like if internet media had matured in the early aughts. How would the memes have read? What would the TikToks say? How fast, how far, would the trutherism spread? Novel lies that inspire fear generate more engagement than truth, a fact that sits uncomfortably next to our platforms' source of revenue: the engagement algorithm. This means that platforms' commercial aims—and the aims of engagement and growth in particular—are at odds with the kind of management that would create an information environment with fewer lies and less fear. Platforms' ravenous appetite for content, their hunger for engagement, their business models' need for growth at all costs, often overwhelms whatever civic impulses the companies might harbor.

In 2017, amid a campaign of state-backed and systemic violence against the Rohingya, a mostly Muslim ethnic minority in Myanmar, Facebook's "dangerous algorithms and reckless pursuit of profit substantially contributed to the atrocities perpetrated" during the campaign of ethnic cleansing, according to a report by Amnesty International. "The display of inflammatory content—including that which advocates hatred, constituting incitement to violence, hostility and discrimination—is an effective way of keeping people on the platform longer," the report reads. "Meta received repeated communications and visits by local civil society activists between 2012 and 2017 when the company was warned that it risked contributing to extreme violence."

Facebook vowed to change. In a 2018 letter from Facebook CEO Mark Zuckerberg to civil society groups in Myanmar, reported by the *New York Times*, he touted efforts to improve Facebook's "technology and tools." The company added additional Burmese-language reviewers, increased head count for Myanmar-related issues, and spooled up a "special product team working to better understand the specific local challenges and build the right tools to help keep people there safe."

But the core business model didn't change. One year later, an internal document from Facebook—eventually leaked to news outlets in the so-called Facebook Files—reported that "hate speech, divisive political speech, and misinformation on Facebook" continued to thrive, and that there was "compelling evidence that our core product mechanics, such as virality, recommendations, and optimizing for engagement, are a significant part of why these types of speech flourish on the platform." And just as the engagement engine didn't stop, Facebook-linked violence didn't stop either. In 2020, internal research found that "Facebook's products in India were awash with inflammatory content that one report linked to deadly religious riots," Newley Purnell and Jeff Horwitz wrote for the *Wall Street Journal*, relying on the leaked documents in the Facebook Files. "Inflammatory content on Facebook spiked 300% above previous levels at times during the months following December 2019, a period in which religious protests swept India." In February 2020, according to the Facebook researchers, "communal violence in Delhi left 53 dead," Purnell and Horwitz continued. In response, Facebook researchers interviewed dozens of users in India. "Most participants from Hindu and Muslim communities felt that they saw a large amount of content that encourages conflict, hatred and violence on Facebook and WhatsApp," the researchers said, noting that most believe it was "Facebook's responsibility to reduce this content." But what would it look like to actually reduce it?

Concerns about the rise of harmful content are often greeted with a non-sequitur defense of free speech. Non sequitur, because the constraint isn't always on one's right to *speak* the falsehoods, but rather, often, on the privilege to have those falsehoods disseminated widely. A misapprehension lurks in the "free speech" arguments against content moderation. When users complain that their speech is being throttled, the question I want to ask is, "Relative to what baseline?" That is, why do users expect their content to be seen and engaged with by so many others? Not every post is going to do numbers.

Recommendation algorithms make all manner of decisions about what kinds of content to spread. Those decisions are, in the end, arbitrary. Platforms have experimented with enshrining those arbitrary decisions into a faux-legalistic regime. Facebook's "Oversight Board," for example, promises to "apply Facebook, Instagram and Threads' content standards in a way that protects freedom of expression and other global human rights standards." Court-like offerings, such as the Oversight Board, offer a patina of fairness. But beneath that patina rests an algorithm that will, necessarily, "moderate" the content that is disseminated. An engagement algorithm will privilege engaging content—with all its novel, fear-inspiring lies—just as an algorithm that privileges time-on-site will privilege content without links.

An eagle-eyed reader might accuse me of constructing a straw man here. After all, content moderation does more than restrict algorithmic reach; it bans and scrubs speech as well. Even more, often the complaint isn't that platforms are choking some broad category of content; it's that platforms are choking out some specific political tilt.

As to the former point, I agree that platforms, at times, have regulated—and do regulate—speech by removing it (and even removing the account of the person who posts it), though I note that platforms have increasingly moved away from hard-and-fast bans of posts and toward limiting the virality of content they deem problematic. But on the latter point, as long as platforms offer one-size-fits-all engagement algorithms to recommend content, we will always be privileging some ideologies over others. Angry extremism engages more than staid moderation; false stridency will beat real nuance every time. The fact remains that the choice of algorithm—the medium, media theorists might say—dictates the messages.

To me, this kind of medium-level moderation matters far more than the specific political features of a given platform's content-level moderation tactics. The trouble is that civic-minded medium-level moderation would likely entail a smaller role for engagement algorithms, but the primacy of those engagement algorithms is exactly what has powered the platforms' growth.

Any fair assessment of the present state of digital media must reckon with its constructedness. Facebook's core product mechanisms were not brought down the mountain on stone tablets. There are no fixed rules to the digital ecosystem; platforms don't interfere with a natural state of being. We must recognize, both as the creators and users of the web, that we live in a system whose rules can be, and in fact always are, constructed. The fantasy of a ruleless, idealized "town square" where anything is fair game remains just that: a fantasy. It tempts, though, the notion of a natural way to be; what, we can't seem to help but wonder, is the natural state of human discourse?

"One way to guarantee a conversation without a conclusion is to ask a group of people what nature is," Rebecca Solnit writes. The reason for all the trouble is that *nature* is an ever-moving target. In Yosemite, Ansel Adams and his cohort of so-called nature photographers found "an untrammeled landscape, virgin and pristine, not a bootprint to be seen, not a hogan or teepee in sight," Mark Dowie writes at *The MIT Press Reader*. "Here in this wild place one may seek and find complete peace." In truth, Yosemite was a managed place. "Adams assiduously avoided photographing any of the local Miwok who were rarely out of his sight as he worked Yosemite Valley," Dowie writes. "He filled thousands of human-free negatives with land he knew the Miwok had tended for at least four thousand years."

In much the same way that our IRL ecosystems always are—and always have been—managed, that humans are active agents in the creation of nature, so too must we recognize that we have a responsibility to manage our digital ecosystems. While we often frame digital land management decisions as a fight between censoriousness and freedom, we would be better off if we remembered that there is no such thing as a "natural" or "neutral" set of rules. Freedom springs from action, agency, and choice, just as censorship does.

We build and manage the web—all of us. We define our own rules every time we tap, click, or scroll. In my first experiment in world-building, when I was a child (and into my early twenties, embarrassingly enough), I created entire cities made of paper. I made little sticky-note origami citizens, meticulously decorated, who lived and worked in buildings made of used printer paper. The capital city, home to hundreds of paper people, each fastidiously named, stretched out over a dozen square feet and grew to four feet tall in places. (I didn't have a lot of friends.)

As I grew older, acting as the god of a little creation became less and less interesting to me. Worse, it took up too much space, and hauling it from apartment to apartment proved a challenge. But online, space is cheap—almost free. Online, with a little education, we can code our own paper cities. Online, we can build without permits, without end. Online, we use the web to practice dying, which is to say, living—truly, in its fullness. How strange, then, that we've entrusted the management of our digital metropolis to others. How strange that we've traded our agency for venture capital–backed advertising business models.

The digital horizon stretches out before us, not virgin and unblemished, but marked everywhere by us. Long-forgotten posts cast their shadows, which stretch and warp across the face of the internet. Wind whips through ruins of websites, the impression of our footprints covering every inch of ground. This may be ours, but it's still untamed. The vast wildness of it might inspire, but it just as easily overwhelms—drives us mad in its rich endlessness. This world spins off its axis, and we control everything and, it seems, nothing about its tilt.

Part 3

BUT THERE IS A TIME TO . . .

There is a time for everything,
and a season for every activity under the heavens:
a time to be born and a time to die,
a time to plant and a time to uproot.

—Ecclesiastes 3:1–2 (NIV)

CHAPTER 10

WALK

The Hudson River flows a mile from my house, down a quiet road that passes several dozen acres of wooded conservation land. In the spring, little wildflowers appear in the ditch—purple, white, and even orange, somehow both delicate and sturdy—my favorite of Mother Earth's magic tricks. A hill rushes down to a gorge with a little creek slicing through it, the same creek that marks the boundary between me and my neighbors. The shadows of the trees, a collection of maples and oaks and the occasional birch, pine, and poplar, mottle the road and hill and gorge with their leaves, which stretch in the breeze that rises off the Hudson. In the warm spring rains, the creek swells, overflows its thin banks, and rushes to the river, which sprints to the ocean, a network of nature that terrifies in its inhuman breadth and power. What a relief to feel awe again. All I need to do is open the door, walk.

Every day this year, even when sheets of rain blanket everything, even in the thick humidity that marked this year's turn to summer, I tried to walk the road down to the little park at the edge of the Hudson. It's hard to remember to forget my phone, and I still don't always succeed. After all, I enjoy taking pictures of particularly striking plants, identifying them with an app's help. Here, the delicate patterning of Queen Anne's lace. There, the droll shape of the jack-in-the-pulpits. Further, the happy sprouts of wild garlic. Down by the creek, the overwhelming perfume of skunk cabbage. Everywhere, the thorny promise of the wineberries and black raspberries. When I forget to forget my phone, I can still eke out awe, but only when I convince the thing to stop sending me pings and vibrations, only when no one wants or needs to get ahold of me. It turns out that when I let myself be disturbed, I often am.

Despite what you may think, I am addicted to my phone. My partner has turned her phone's screen grayscale, uses a "brick" to lobotomize many of her applications, and regularly removes social media and email. Even contemplating these measures gives me palpitations. Even a phantom buzz is enough for me to dive my hand into my pocket and retrieve my phone, expecting what I don't know. On my phone, I listen to podcasts with hosts who smugly tell me how they've kicked their smartphone habit. On my phone, I read articles with authors who joyfully proclaim the gospel of a slow internet. On my phone, I watch videos advertising products that promise to heal me. This doesn't work.

And still, I am drawn—as we all are—to resist in some way. I am desperate to push through the algorithms' currents, through the unceasing flow of notifications, to scrabble up a rock and take a luxurious breath. Nature works, though awe is rarely practical.

I know it's gauche to say it, but I love my work, even if I don't always love my job. I find meaning in it, in adding to the storehouse of public knowledge. And more often than not, my work happens in the digital world, in the denuded attentional world, full only with buzzes and pings. What does resistance look like there? Every once in a while, in my scrolling, something appears that makes me laugh, and that laughter, more than anything else, offers a chance to pause in the rushing stream. When this happens, I often memorialize the occasion with a screenshot. This means that, for every social media platform I have an account with, I likely have a screenshot (or several) squirreled away on my computer, in a folder full of other posts that made me laugh.

Like this one, a tweet by the journalist Eve Peyser, posted in November 2017, one of hundreds in a folder on my desktop: "Now, this is a story all about how / My breakfast plans turned upside down," it reads. If you, like me, find this joke uproariously funny, then congratulations, you were also far too online in the late teens. If you don't (bless you), then forgive me for what follows.

Peyser's joke splices two species of meme—two strands of content, repeated again and again with variations, reproducing through imitation, passed from person to person, accumulating meaning with each transmission. The first meme: Twitter's inexplicable remixing of William Carlos Williams's poem "This Is Just to Say." The original begins: "I have eaten / the plums." For whatever reason, users of Twitter have delighted in tweaking the words of this poem, bending its meaning toward ever more absurd ends. The second meme: various riffs on the theme of the television show *The Fresh Prince of Bel-Air*. The opening words of the song read: "Now, this is a story all about how / my life got turned, flipped upside down." Peyser nimbly mixes the two together, offering a joke about these jokes. And since explicating a joke makes it funnier, I'm sure it's funny to you now too.

I don't remember what I was doing exactly when I saw Peyser's tweet, though I can guess that I was midstream, paddling futilely against the current. And I paused. And I took a screenshot. Delight is awe's twin in wonder. These siblings wake me from my glassy-eyed stupor in a way little else can. Together, we explore the internet at its best, a place of writhing, bacchic humanity. Too often, the web's algorithms and pages sleepwalk me from next to next to next.

I don't want to put too much weight on Peyser's little joke, but I set time aside when I laugh, and in so doing, I take a sabbath. The Sabbath, a day set aside, is "a day of sacred assembly," Leviticus 23:3 (NIV) tells us. In other words, Hebrew law locates the holy inside the use of time. When we call social media a stream, we are naming it as a place where we find both the base and the sacred, a place where a sabbath might be in order. I have a pet theory that everyone, religious or not, finds their own way to a kind of sabbath. After all, human culture, just like nature, can be red in tooth and claw; for a practice like the Sabbath to have survived this long means it has proved useful and adaptive to generations of people. This kind of cultural practice, passed down through the ages, can be helpful for believers and nonbelievers alike. Recognizing a sabbath doesn't have to mean becoming religious. Whether or not you think God demands it, the Sabbath allows us to metabolize our own mortality.

The base unit of a social media post isn't really the single, atomized post; it's the conversation in which we have embedded it. When I capture one of these lonely screenshots from X or Bluesky, I have missed the posts before and after, the replies from other users—the zeitgeist of the moment. You wouldn't know any of this from my screenshot. You wouldn't know the long tradition, the hundreds of posts, that make Peyser's joke work, that made my Sabbath worth taking. With each screenshot, that context is lost—lost forever, since an algorithm decides which posts to give me at any given moment. This point is crucial: Our feeds are likely unique and cannot be reconstructed. Neither the king's horses nor the king's men could say what tweets I saw in the order I saw them a week ago—to say nothing of a year ago. But my own personal Humpty-Dumpty is what I want to save: the conversation—so fragile!—I witnessed and added to.

It isn't only jokes that work this way online. A tweet is a social activity. Each one ties together a thousand other tweets, responds to its author's ephemeral feed. In May 2021, I posted link after link about the Tulsa Massacre, a time one hundred years before, when a white mob razed a prosperous Black neighborhood to the ground. These posts were in conversation with my feed, with a commemoration of a historical event and the discourse that had sprung up around it. I saved a screenshot or two from this time, links to the *New York Times*' re-creation of Greenwood—the neighborhood that was burned down—as well as some posts about the series of stories about Tulsa that I worked on for the *Wall Street Journal*. These screenshots point back to that moment in time: a conversation about racism and violence anchored in the aftermath of the protests over George Floyd's murder. But though these tweets still have meaning absent their 2021 contexts, they don't capture that moment. They lack the urgency of that time, the heat of it.

Since I also mark up my books—underline them, add marginalia, and even sometimes save a picture of a page to the same folder where I save social media posts—you'd be forgiven for thinking not much has changed. In one way, you'd be correct: The paragraph is no more the unit of a book than a tweet is the unit of a conversation. But unlike a book, where I am referencing something fixed, something that can be re-created in its wholeness, an online conversation fuzzes at the edges, loses meaning even as I grasp for it. Digital media is a school of fish, a swarm of locusts, a pointillist painting. It is an object made of smoke. Taken at a distance, its facade seems solid, but chisel out a brick, and you're left with sand. The web loses coherence with every day that passes into days, months, and years.

And thank God for that! I've cited it before, but "time is a privacy setting," as John Herrman declared at *The Awl*. "A single user's Twitter archive is a series of permanent and public contributions to discussions that have long since ended," he wrote. "In the archives . . . each speaker is isolated to the point of incoherence." But there's more than incoherence lurking in the lossy format of the internet; there's danger too. Wrested from their contexts, stripped from their times and places, our isolated words can turn on their owners. Jill Lepore noted at the *New Yorker*, "It might seem, and it often feels, as though stuff on the Web lasts forever, for better and frequently for worse: the embarrassing photograph, the regretted blog." Context decays. A tweet is an elegy to what it signifies.

What is the point of an elegy if nothing dies? The Sabbath requires loss, requires danger. Absent forgetting, absent idleness, it would have no meaning. The Sabbath doesn't eliminate choice, but it does reclaim it: I choose to forget; the platforms don't choose for me.

CHAPTER 11
READ

By parental fiat, when I was growing up in Minneapolis, I couldn't ride my bicycle beyond Bryant Avenue to the east, Kenwood Parkway to the north and west, and Lake Street to the south. At first, I didn't complain; the East Isles neighborhood of my childhood was enough for me. But as adolescence dawned, the constraints chafed. So, as a tween, when the restrictions were lifted, when I could bike happily around all of South Minneapolis, I did just that—exploring every nook and cranny of the city like it was a feast. It didn't take long to discover the Walker Art Center's Sculpture Garden.

Famed for its Claes Oldenburg sculpture *Spoonbridge and Cherry*—a giant spoon that arcs over a little pond with a cherry on the end—I used to spend sun-drenched hours jumping from piece to piece, admiring each installation in turn. The sculpture garden is bounded on one side by the coming together of Hennepin and Lyndale Avenues and Interstate 94. Across the sixteen lanes of traffic sits Loring Park and then downtown Minneapolis. To cross the mess of cars, you can climb the steps up the artist Siah Armajani's *Irene Hixon Whitney Bridge*—a baby blue– and pastel yellow–painted splash of steel and air with picture-perfect views of the downtown skyline. Armajani inscribed on that bridge the first poem I ever memorized, one without a title, written for the bridge by John Ashbery. For the Walker Art Center, Paul Schmelzer noted that, as the poem covers 375 feet, the museum staff "have joked that the untitled work is the longest poem in the world."

In high school, crossing the bridge for the first time, I looked up and read the poem's words shouting down at me from their raised, all-caps golden inscription. "IT IS SO MUCH LIKE A BEACH AFTER ALL," Ashbery writes, "WHERE YOU STAND AND THINK OF GOING NO FURTHER. AND IT IS GOOD WHEN YOU GET TO NO FURTHER. IT IS LIKE A REASON THAT PICKS YOU UP AND PLACES YOU WHERE YOU ALWAYS WANTED TO BE." On that bridge for the first time, I realized that this was indeed where I had always wanted to be—a poem's words etching themselves into my synapses, inscribing themselves there long after I left the poem's presence.

In Garth Greenwell's book *Small Rain*, the poet who narrates the novel endures a claustrophobic MRI experience in immense pain. To distract himself, he recounts George Oppen's poem "Stranger's Child" again and again. Though it's fewer than a dozen lines long, Greenwell explicates the poem for nearly an entire chapter of his book, deep in the mind of his unnamed protagonist. In the novel, it works—or sort of works; neither the pain nor the horror of the experience lessens, really, but while his body tenses in the machine, he dwells somewhere else too. He rests inside Oppen's words as well as his own mind.

I've noticed my own fascination with the recurring paradoxes of the human experience. We refresh attention, I've argued, as we use it. We unlock the benefits of reading, I've claimed, only when we practice it noninstrumentally.

And here's another: Greenwell's narrator finds ever more space as he narrows himself into the confines of Oppen's words. Greenwell, in an essay in his newsletter *To a Green Thought*, explores "how memorizing poems will change your life." The goal of this memorization, too, is a paradox. It at once gathers and knits up our consciousness, Greenwell writes. It is "an expansion of the self but also a drawing of the self to a point." The arguments I've been putting forward, in other words, mirror the strangeness of the internet in all its paradoxes and wildness.

From the outside, it's hard to see a difference between, on the one hand, mindlessly scrolling on your phone and, on the other, mindfully doing the same. It's hard, really, to point to what about the physical book (or ebook reader) makes *reading before bed* a different kind of activity than *scrolling before bed*. But those activities certainly feel different while I'm engaging in them. Greenwell feels the difference too. Like so many of us, he sees how the experience of social media "scatters consciousness and so diminishes it." The practice of poetry, he says, is "the best treatment I know for an addiction to doomscrolling."

I can still recite the untitled Ashbery poem from my youth, and I have recited it—sometimes in the same way Greenwell's narrator did—as a kind of defense against the world's attentional predation. And pain, more than any other sensation, reigns as king of the attentional jungle. What is pain, after all, but the hijacking of your mind? Whether physical or psychic, pain, to quote one of John Green's characters in *The Fault in Our Stars*, "demands to be felt." When we recite a poem, we are calling forward a memory of the past, guiding it into a new orbit, where it can eclipse the pain of the present.

When we remember a poem, when we, say, memorize it so we can recite it in times of duress, we allow our attention to gather and spread inside a medium we control. We reclaim agency as we remember another's words. The digital world is full of memory aids. We pull up maps on our phone to get from our apartments to the nearest grocery store. We take notes from a meeting and store them on our tablets. We bookmark our favorite websites in the browsers on our computers. All these rest inside a medium we control far less than our own minds. They are subject to rot and retention based on what some third party thinks is essential, not our own choices. What happens when Google Maps, after being paid by some business, advertises an ice cream stand between you and your grocery store? What happens when the note-taking app runs out of seed capital and shuts down, leaving your notes stranded in the cloud? When happens when the vagaries of internet life change the sites we bookmarked so carefully? Of course, humans forget things all the time; our brains, fragile and fickle, often need to forget in order for us to operate in the world successfully. Yes, time does heal all wounds, even the emotional ones, in part by letting us forget the pain that accompanied them. But these little forgettings are *ours* in a way a website never can be.

I'm sorry for the brag, but I used to be a world-class recorder player. I studied baroque music—a period when the recorder was more than a curiosity or a starter instrument for children—in conservatory and played in competitions, where I did very well, thank you very much, and I even won some awards for my playing. But I have forgotten, as my career drifted away from music, how to play the instrument I love. I can still remember the notes, but much of the musicality and fluency have left me. While this forgetting saddens me from time to time, I chose it. And I recognize that the choice to forget was required by the career I have now. Time, after all, is finite. The internet's forgetting of *Ich Bin Ein Oberliner*, on the other hand, I did not choose. Both losses make me sad. Only one frustrates me.

Remembering a poem, then, reclaims our attention for ourselves because it takes place in a medium we mostly control. Memory, as I've argued, is both social and physical. Social because it requires institutions (as in the case of our remembering of Bach's music), and physical because it takes place inside our bodies (as in the case of our recitations of Homer's *Iliad*). It's even there in the word, *to remember*, which contains the idea of a *member*, a component of a body, of a limb. *To member* something would be to place a sensation, an experience, or a thought into a corporeal form, and *to remember*, in that sense, means to give an idea flesh once again. We have a phrase, *muscle memory*, that we deploy around physical skills—when an activity is so deeply ingrained in us that it doesn't seem to originate in the brain.

An old friend of mine, a piano prodigy, used to remember phone numbers (we used to have to remember phone numbers!) by assigning each number a note on the piano. He'd imagine "playing" the phone numbers on the keyboard, and the shape and sound of each number, the melody it made, would help him remember how to call me. Now *that* is muscle memory. These days, when I pick up the recorder, I feel my leaden fingers struggle to make the trills the way they used to. Muscle memory, muscle forgetfulness—our bodies are the members where we store our lives. Their silicon analogues, housed in those power- and water-hungry warehouses, aren't ours, so offloading our memories there entails great risk.

Here's a metaphor: During the COVID-19 pandemic, many Americans, including me, were shocked to discover our dependence on goods produced overseas. Producing even something as simple as a mask required far-flung factories, so our impoverished domestic production capacity became something of a liability. In normal times, cheap goods could flow across oceans and flood big-box stores around the country. Globalization usefully gives us inexpensive televisions, computers, clothes, and more. But sometimes you need to produce something closer to home. And if you lack the facilities to produce, if the tools and machines your factories need have moved elsewhere, it will take time (and economic pain in the meantime) to rebuild. When we're doing well, in an attentional sense, remembering a poem might be a luxury. But when we need our memories back, that work might be essential.

And poetry offers more than just a flex for our brains. Poetry, the act of reading it, requires a different kind of attention and care than reading other kinds of texts, or engaging in other kinds of activities. When we call poetry "difficult" to understand or to read, I think we're pointing at the difficulty of entering this other mode of reading, a mode that demands more—and differently—from our attention.

Dead-tree books persist—thrive, even—in this golden age of phone and computer screens, of e-ink displays and iPads, because at some level we hunger for interfaces that encourage close reading. In the same way that a starving man develops particular hungers for the nutrients he requires to survive, a reader develops a taste for physical pages that bound and shape a reading experience—she needs it "like a thirst for salt," to steal Robert Hass's metaphor from his poem "Meditation at Lagunitas." One academic study, in the *Journal of Educational Psychology*, found that students retained more from words printed on paper than from words pixeled onto a screen—a well-replicated finding. But the same study found that "handheld devices may still play a role in supporting students' text comprehension. As compared to computers, handheld devices . . . allow physical interaction that is closer to reading in print." The authors found that e-readers allow their users to "get physically closer to the text" and that, since the handheld device replicates the layout of a printed page, they allow "strategic backtracking," letting their users move back and forth like they would in a printed book. Screens are inferior to the pulp-and-ink book: That's true. But not all screens are the same. What we bring to the screens matters. How the screens present text matters. And what we do with the text matters even more.

Since reading Greenwell's essay, I've started trying to memorize poetry again, now on my computer screen. While I have plenty of books of poetry in my library, I signed up for a newsletter that provides a poem a day, and every few days, when I discover one at the center of a Venn diagram of *short enough to remember* and *speaks to me*, I try to memorize it. Having the poem pop up in an inbox full of important tasks (and idiotic trifles), offering something different, helps keep my head on straight these days. Though I'm out of practice, there's real heat there, as I rub my mind against a poem's words, the friction of it (yes, that word again) throwing off sparks of meaning. I started where Greenwell's narrator in *Small Rain* ended, with Oppen's poem "Stranger's Child." It concludes: "The sparrow's feet, / Feet of the sparrow's child touch / Naked rock." A few recitations in, I started noticing the repetition of the word *feet*, the way that repetition drew my attention down to the naked rock those feet rest on. Maybe it's too cute to think of the sparrow's feet as the poem's very language—*feet* can also refer to a poem's base unit of meter—resting on some kind of bedrock of meaning. Maybe that's an over-reading of Oppen, something I'm bringing to the text, not something already in it. Maybe, but so what if it is? I was delighted by this discovery—or creation. I felt charged with it. I wouldn't have gotten there if I hadn't been trying to remember the poem.

CHAPTER 12
DREAM

I learned to read poetry in Michael Bazzett's high school English class. At the time, I didn't realize that Mr. Bazzett was an accomplished poet in his own right, though I've since read several of his books—including his luminous translation of the Mayan epic *Popol Vuh*. A somewhat surly, broken teenager, I was never a good student. It wasn't that I slacked off—not exactly, anyway—but the things that caught my attention were rarely useful for grades. Mr. Bazzett had the gift of seeing my interests, and the interests of all the others in the class, and twining the curriculum with those interests, bending our attentions toward educational purposes. It was in his class that I first read serious contemporary poetry seriously. We read, among other books, Robert Hass's *Praise*, which "Meditation at Lagunitas" is from. I would look out the grand south-facing windows in his classroom—where he still teaches—and I would struggle and sometimes succeed in drawing meaning from poems' deep wells, the sunlight brightening everything, even the dark recesses of the poems we discussed.

The occasional draft of understanding slakes more than thirst, it turns out. "A poem is at the very least a record of one's survival," Billy-Ray Belcourt has his narrator say in his short story "Poetry Class," collected in his book *Coexistence*. I've loved this line since I read it because it points to the kind of engagement poetry demands. Belcourt's words are true in an obvious sense—a poem couldn't exist absent a writer and a reader, after all. And in context, his narrator is tying the practice of poetry to Indigenous resistance of colonialism—a point well taken. But if you'll indulge some obsessing: The word *survival* also suggests other meanings. Belcourt, after all, could have written, "a record of one's *life*." But instead, Belcourt gives us *survival*, which gets at something deeper: It makes a poem's record central to human experience, as necessary as food, water, or shelter. Every time we read a poem, recite it aloud or in our minds, we are blowing, gently, on the flame of this other way of being, keeping it alive. Yes, *to survive* can mean, thinly, *to continue living*, but as Jesus says in the Gospel of Matthew says, "Man shall not live by bread alone" (Matt. 4:4 KJV).

In the last paragraph, I conducted what Mr. Bazzett might have called a "close reading" of a sentence of Belcourt's prose. This kind of reading, the trick of it, the same kind of reading a poem encourages, rests on the assumption of human meaning behind the words. For the practice to mean something, we must assume some kind of intent in there. Some theories dismiss this assumption as hopelessly naive—and maybe it is—but I'm no theorist, and I've found it heartening to imagine that as I read and negotiate with a text to pull out meaning, I'm in conversation with its authors and their aims.

Of course, online, a close reading can feel foolhardy. A dashed-off Instagram caption, a brief Facebook post—are these texts worthy of the same kind of reading as a line of poetry or a lyrically written sentence in a book? Even more, these platforms are now rife with the outputs of artificial intelligence, and these large language models can seem to produce, in the *Atlantic* writer Franklin Foer's prescient and memorable phrase (which titled his book on the web), a "world without mind."

A close read of an AI-generated text, to my humanist side, feels almost sacrilegious. Sacrilegious—and boring. Is the future of platforms mostly automated accounts producing AI slop, which is then sifted by a different AI in order to be served to other mostly automated accounts? Close reading seems utterly pointless when you're delving (as an AI might say) into a bulleted and emojied summary of some other text produced by AI, when we're all just "inflating and compressing content generated by A.I.," as Ezra Klein quotes one computer scientist in the *New York Times*.

Two questions, then: Can a computer aid, really, in a close reading? And in the age of computer-composed texts, is close reading worth it? There are at least two distinct (though related) threats, then, to close reading in our digital world. First, the platforms on which internet texts appear have not been designed to encourage close reading, and second, the platforms on which internet texts appear are deploying large language models in order to create and read texts for you.

In 2017, long before OpenAI released its internet-breaking large language model, ChatGPT, I constructed a very little language model with one mission: to produce authentic-ish prose in the style of Herman Melville's *Bartleby, the Scrivener*. To train my model, I split up *Bartleby* into 140-character chunks. Then I asked my model to predict what the 141st character should be. Through a fanciful-sounding technique called "backpropagation through time," it would be nudged toward more likely next characters and away from less likely next characters. Every few rounds, I had the model write some faux-Melville for me, with the starting text "I would." After the first fifty passes, it had figured out that it ought to put spaces between chunks of letters. "I would doen mesing thand," it garbled. After five hundred passes, it knew how to spell. "I would jail I think," it wrote. At around five thousand passes, it almost got it, declaring, "I would prefer to, yes." I recently resurrected my now-antique code, remembering, a little wistfully, the excitement that had swept through me as it learned to write like Melville—a species of the pride I felt when my daughter learned to speak.

But I wasn't feeling only pride; I felt terror too. If I could do this with my laptop, what could the Silicon Valley sorcerers with their supercomputers conjure up? We are now in full artificial-intelligence fever, burning with questions about speculative bubbles and even more speculative apocalypses. AI is coming—is already here. Even if the most bullish predictions of a general kind of superintelligence prove false (as I think they might), we'll need to live with this eerie and powerful creation. But as long as we cannot touch these models, as long as we cannot roll up our sleeves and put our hands on their humming bodies, our humanity will be at risk. Tech companies present AI to us in the form of seamless chatbots that spit out everything we could possibly want to know. But this is deceptive. Meaning takes effort. Creativity requires friction. Purpose emerges when we rub our expansive wills against our finite abilities. This is what Mr. Bazzett taught me.

The library in Jorge Luis Borges's short story "The Library of Babel" contains every combination of letters, spaces, and periods in 410-page books of 1,320,000 characters. It contains every novel and religious tract, every fiery manifesto and unanswered love letter. It also contains book after book of meaningless nothing. Naturally, the librarians are driven insane. The academic and technologist Jonathan Basile created a website of the same name that contains all possible 3,200-character books—and a search bar. Made possible through some computational wizardry, Basile's library is equal parts addictive and unsettling. Idly curious, I searched "this is how john west will die." Pages of results sprung into existence. On one, the next word read, "appendix." On another, "catastrophe." On a third, "lust." I could keep going, I realized, and find every word in the English language as the cause of my death. One of them is surely right.

It took me seven years to write my first book, *Lessons and Carols*, but when I entered my opening paragraph into Basile's library, it was already there, just waiting to be discovered. I understood, in that moment, the surprising smallness of human endeavors. Love, joy, the riches of humanity—they all existed here, but they meant absolutely nothing. Borges never saw the rise of AI language models, but "The Library of Babel" presaged them. Just as every piece of language is in Borges's library, waiting for a human reader, so, too, is every arrangement of words and meaning somewhere in our machine models, waiting to be discovered. Borges's old story offers a reminder for our artificial intelligence present. If everything is available, nothing has meaning. Remember your finitude, it suggests, because friction safeguards our sanities.

It might be useful here to know a bit about how these large language models work. A few years ago, I worked with my partner, the artist and writer Galen Beebe, to analyze her first kisses with everyone she ever kissed. In *Kiss List*, I use numbers to represent each kisser's astrological sign, their gender, how far they went after the kiss, how much they talked afterward, how much Galen enjoyed it, how much shame she felt, and so on. Each number is one dimension of something called a vector, and each vector represents a kiss. Think of a vector as a point on a piece of graph paper. That point can be identified with just two numbers, one plotted horizontally and the other vertically. One number gives you some information about one thing, and the other gives you information about another. Add an axis coming out of the page, and that number is a third dimension. Color-code the points, and you have a fourth. Large language models employ the same concept—the vector—though with many hundreds of dimensions, not two or three or four. It is through splitting the human world into long sequences of numbers, into these vectors, that algorithms are able to understand it.

Take these lines from Hass's "Meditation at Lagunitas": "Because there is in this world no one thing / to which the bramble of blackberry corresponds, / a word is elegy to what it signifies." A (very simplified) language model would "read" these lines by hiding *because* and working its way to *signifies* again and again, trying to predict, each time, what word is hidden. After each pass, the model assesses how far off its prediction was, tweaks its machinery, and tries again. This model would contain layers of hundreds (or even thousands) of "neurons," each a mathematical function.

Each neuron carries some number of "weights"—numbers representing its state. As the model hides and guesses again and again, each neuron adjusts its weights, which are nudged so that every guess is a bit more educated than the last. In the end, after reading many more poems, we can extract the outputs from the last layer of neurons, which become our vectors. And just like how we can plot a distance between two points in a scatter plot, the relationships between a model's vectors are plottable as well. In short, if a model reads enough poetry, it would know that *blackberry* and *bramble* are related—that their vectors are similar—but that *blackberry* and *because* are not.

You can do math with these vectors. These words relate to one another in a "vector space." Like stars in a galaxy, they can be tightly packed next to one another, or they can languish alone in the dark expanse. Distance is an easily computable property. Google researchers discovered that *king* minus *man* plus *woman* was tantalizingly close to *queen*. In other words, meaning, to an algorithm, can be added up and averaged, divided and subtracted.

In a different blackberry poem, "Blackberry-Picking" by the Irish writer Seamus Heaney, the speaker hoards berries in a cow barn, but "once off the bush / The fruit fermented, the sweet flesh would turn sour." Here is this line after I nudged each word's vector toward the vector for *angry*: "prompting off angry angered / Angry vegetables fermented, angry bitter flesh prompting enraged sour." *Fruit* transformed into *vegetables*. *Sweet* turned *bitter*. Filler words—*once* and *would*—became, inexplicably, *prompting*. When we find that, to a model, an angry fruit is really a vegetable, we learn something about that model. But more importantly, we're also learning about the meanings encoded into all the texts it's trained on—in the case of large language models, that's pretty much everything. AIs have all the texts we've ever put online, which must be a significant portion of all the texts we've ever written.

It's not only words that can be represented this way; sentences, paragraphs, entire documents—all can be encoded as vectors.

Here I asked a more sophisticated model to find the midpoint between the Hass line "a word is elegy to what it signifies" and that Heaney line about blackberries, and produce a sentence with it. After futzing with punctuation for legibility, this is what it came up with: "Once upon a time, there was a time. Once upon an era. Once on earth. Once in the sky. Once, once." With these passages, I've been pushing against the constraints of AI systems, and the result (if I do say so myself) is something meaningful—beautiful, even. AI can unlock a liminal space, once confined to poets, allowing anyone to explore its dimensions of meaning in new ways.

Throughout my AI work, I've generally operated with vectors—not their slick chatbot facades. To work this way, you need to know how to code. But that needn't always be the case. Our AI companies could create interfaces that allow nontechnical users to operate AI systems in deeper and more meaningful ways. And my Melville model offers another lesson: Its source is clear—and in the public domain. OpenAI and most other AI companies won't reveal what their sources are. They mask the human labor that stands behind their models, and in so doing, they alienate us from ourselves.

ChatGPT's ease has its utility, yes. As a software engineer in the newsroom, it helps me think through the structure of my code, helps me remember the right functions to use. But that utility has its costs: We are missing systems that more transparently reference their training data, letting us better understand how and why our models are returning what they do. We have taken the collective labor of generations and made it an "other"—a servant to do our bidding. But that other is just us, hiding behind the gauzy scrim of a web interface. My hope for these technologies is that, despite their incomprehensibility and inhumanity, we might still make meaning from the vectors they make of us.

Even before large language models, the world refused to yield its meaning easily. We have always needed to harvest it. AI has only made our task that much more obvious. The blackberries we seek to pick remain on the cusp of rot, and we still have to pick them. We have always lived in a Library of Babel, struggling to find the book that will explain it all.

Speaking of Borges, the artist Marcus Merritt posted a little joke on Bluesky that I can't stop thinking about. "A Borges story about a guy who gets AI to summarize all the world's information for him, and then summarize the summary, until the AI has the whole world summarized into a single word," he writes. "He sits alone at his desk, staring at the word, repeating it endlessly, certain he is experiencing everything." Reading summaries of summaries—this is the power of the large language model. They summarize the world every time we query them, traversing the vectors of human creativity like a kind of intergalactic explorer. "Compressing the entire internet into interpolable vector space that can be queried in natural language remains actual sorcery," Bluesky user @tripsnek noted. But this magic can be put toward meaningful uses and less meaningful ones. We skipped past the power of these models, that user noted, "because we immediately wanted to know if it could write f'in poetry." It can write poetry, it turns out. I'd still rather read humans'.

AI's boosters say the technology will replace vast swaths of humanity's creative and mental labor. Speaking personally, I've found a lot of utility in large language models. At work, I use one nearly every day. They excel at pulling structured data out of long swaths of text, something my job often demands. They can summarize groups of long texts, and in so doing find relevant needles in the haystack of a cache of documents. I find that skeptical arguments about AI rarely take into account the deep utility these models offer, not as a replacement for reading and understanding, but as a way of ordering where to start reading and start understanding. Even more than that, we've taught a monkey, finally and almost literally, to compose Shakespeare—an impressive trick.

Complaints that AI is not fulfilling its creators' admittedly far-fetched fantasies (and might never fulfill their breathless predictions of its potential) strike me as overly defensive. Like, yes, annoying people are saying annoying things about these models, but that doesn't mean the technology is worthless. None of this is to say that the AI slop future, one in which human prose is fully, or almost fully, replaced by artificial garbage, doesn't concern me; I, for one, have little interest in close readings of a chatbot's tokens, absent a human's collaboration.

This is where cultural practices matter. I wrote, earlier, that movements aimed at a new technology—moral panics, in other words—succeed when we spin up cultural practices around those technologies. Right now, society is negotiating what these practices will be. Will AI companies need to pay for the material on which they train their models? Will the Sam Altmans of the world succeed in offloading human decisions to a machine? Will we use them for tasks humans can't do on their own, or will we, lazily, let them eat into activities we cherish? What will the norms be around writing a book or an essay or even a poem? I don't know the answer to these questions, but I've come to see my use of these technologies as a kind of vote for the norms I want to see. So I try to avoid replacing my thinking with a machine's, even as I use these models to help me parse documents and extract meaning.

CHAPTER 13
END

In September 1708, while Johann Sebastian Bach composed his cantata *Actus tragicus*, Adolph Strecker, the former mayor of Mühlhausen, Germany, was lying on his deathbed reading *ars moriendi*—literature on the art of dying. Bach's cantata, likely composed for Strecker's funeral, remains one of my all-time favorite pieces of music, one I've had the pleasure of playing a couple of times. In it, Bach is trying to reconcile a paradox. How is it, as the piece's libretto suggests, that "we all must die" and that "we all live forever in Christ"? How is it that "death has become my sleep" and also "you will be with me in paradise"? The tenor's aria in the first half of the piece cites Psalm 90: "Teach us to number our days, / that we may gain a heart of wisdom."

At the center of the piece is a perfect aria. While the violas da gamba (think guitars played like cellos) and the organ prowl in a menacing, steady rhythm, the bass, tenor, and alto voices—the lower three—craft an intricate fugue, each taking turns twisting a dark theme over the others. "Es ist der alte Bund: Mensch, du mußt sterben!" they sing—a passage from Ecclesiasticus, a book of the Bible accepted only by some Christian denominations as canonical. The King James Version translates it as "for the covenant from the beginning is, Thou shalt die the death." Then the soprano enters. "Ja, komm, Herr Jesu, komm!" she sings from Revelation 22:20. ("Yes, come, Lord Jesus, come!") Again and again she implores, begs for the divine, while the lower voices mourn our inevitable deaths. In the end, the soprano silences even the organ. The voice concludes the movement with "tentative optimism and a distinct lack of completion," the musicologist Wendy Heller writes. "Time and tonal movement are briefly suspended." Even that idea, a brief suspension of time, carries a paradox. How can time be suspended *briefly*? Surely, if it were suspended, we wouldn't be able to use temporal words to describe it? And yet.

The ancient Greeks had two words for time: *chronos* and *kairos*. While chronos falls to us mortals, the time ticked away by a clock, kairos represents the perfect moment of action, of decisive beauty. It means godly time. I suggest Bach solves one riddle—how can humans both die and live forever?—by appealing to another: Bach suspends time, transforming chronos into kairos. I sometimes think about chronos and kairos while the metronome of my days moves me from one activity to the next: morning routines and school drop-offs to meetings at work to school pick-ups and dinner prep to bedtime. I find myself grasping for the moments when time moves like honey, when time opens like a bud, revealing gold-dusted petals. That flowers must fade matters little to the insects that live only a season. Perhaps these moments—entire lives lived in a single moment—matter precisely because they pass.

Maybe chronos and kairos require each other in the same way those ancient philosophers claimed virtue required hardship. After all, why be generous when there is no scarcity? Why be courageous when there is nothing to confront? John Calvin wrote in the *Institutes of the Christian Religion* that "as our abasement is His exaltation, so also the confession of our lowliness always has His mercy near as a remedy." But if joy requires pain, fuck that. It might be that the rancid conditions of our digital lives make its respites all the sweeter, but there is no glamour in suffering. In Psalm 42, as a deer pants after water, the psalmist's soul thirsts for God. But that doesn't mean I think God advocates dying of dehydration. Chronos might make us yearn for kairos, but we can have a better world: We are called not to suffer but to work to end suffering. Here and now.

Karl Marx described a utopia in which we would hunt in the morning, fish in the afternoon, graze cattle in the evening, and philosophize after dinner. Marx locates the miraculous—human liberation—inside the deceptive ordinariness of work, which tracks Genesis at least. Through painful toil, Genesis 3:17 tells us, we will eat of the ground all the days of our lives. Maybe that's true even in heaven. Maybe, even there, joy and obligation remain conjoined, as they are here on earth. Yes, our obligations sprout from the vagaries of this life. And through our compost, may the soil grow less toxic day by day, season by season, harvest by harvest.

There's a glib kind of politics that demands we reorder society before any other improvements can be made—glib, but tempting, since American capitalism, racism, patriarchy, and so forth, operating at the societal level, prevent so much human flourishing. Proponents of this argument are correct that these oppressive systems operate at vast scales, and that individual resistance cannot overcome them in the way that, say, a solidarity-based group resistance could. But this true insight can also seem to imply that individual action doesn't matter or isn't worth taking. "There's no ethical consumption under capitalism," someone might say as they purchase a good produced in the worst kinds of ways. Individual action is constrained and limited and puts different moral commitments at odds with each other, the argument goes. So why bother? This is the voice in my head as I write about memorizing poems to combat surveillance capitalism. It won't, alone, prevent rapacious attentional exploitation. So why bother?

I bother because memorizing George Oppen let me inhabit a different way of being a little bit longer. It let me imagine a different world. I don't own any books by Oppen. I don't know which book collects that poem. The screen, so often vampiric, allowed me to find that poem—and many more besides—but I had to take the next step, to memorize the thing, a kind of transfusion of meaning into a bloodless world.

I bother to store my screenshots, and lose the conversation, just as I bother to trawl Spotify for the full album even when it wants me to think only of the song. I push against the world and am satisfied by its hardness, by the equal and opposite force it yields. A richer life is one spent caring. The full album, the remembered conversation, their sounds replicating themselves on my eardrums, those vibrations making meaning in my mind, reconstituting themselves as memories and emotions—this is the inefficient, unscalable ritual that defies algorithms' promises of frictionless abundance.

I've been working on a little internet-based conceptual art piece masquerading as a social network, called *Homer*. Every day, its users may post only a handful of times. Those posts are shared with each user's followers, but they are visible only for the day of their posting. On the hour, a large language model summarizes each post, distilling it down. Each subsequent hour, the summaries are further summarized, reduced in length until only a single word remains. While, maybe obviously, this system wouldn't "scale," to use the language of contemporary social media companies, it does offer something else: a simultaneous sense of urgency and slowness—urgency because the words are fleeting, and slowness because of the app's constraints. I won't pretend that *Homer* will be the ticket out of this dump, but I hope, as with any piece of art, it might push us toward a different kind of imagination, one that centers the human experience in all its paradoxical urgency and expansiveness. But *Homer* won't scale. Neither will Oppen. They are useful but by no means sufficient.

It tempts, the idea that we can sabbath our way out of the web that ensnares us. But I remain skeptical that lone human action, even undertaken in aggregate, will let us escape. Still, it is a start. Though the platforms and the paradigms they work within purport to offer meaningful engagement with the world, meaning requires a different relationship to time, one governed by kairos. Simply put, we cannot measure value in time spent. The bargain the platforms offer, Faustian in its way, trades the Sabbath for the scroll. They'll handle the choice, if only we offer up the numinous power located in human agency. We practice dying every time we use the web, even if we close our eyes and try to ignore our finitude. We can choose to pass our time in the sun-bright awareness of our own mortalities, can take the Sabbath and trade the saccharine taste of an undying internet for the honied amber—slow and sticky—of a full life lived in meaning.

ACKNOWLEDGMENTS

A book is a block of wood touched until it's smooth, the poet Lauren Clark once told me. More hands than mine touched this ragged chunk.

My editor, Lisa Ann Cockrel, believed in my vision for this book, and I will always be grateful for her trust—if a little confused by it. Laurel Draper transformed this from a manuscript into a book, despite every deadline I missed. Victoria Jones has sharper eyes than anyone I've met, even if she was wrong about that one comma. Hilary McClellen remains the best fact-checker this side of the Mississippi—and she checked this paragraph, so you know it's true. My neuroses and poor email etiquette somehow haven't convinced my agent, Zeynep Sen, to fire me.

My first and best reader is Galen Beebe, of course. Her support, on and off the page, made this book possible. My daughter, Lavinia Beebe-West, even without being able to read, touched every word. David Ohana's insatiable curiosity and constant conversations inspired me to keep writing. Gabe Isman taught me to code.

The Lotic School of writers, including the aforementioned Galen Beebe as well as Eugenie Dalland, Kelly Marages, Zachary Spence, and Amelia Theriault-Brown, workshopped every bit of this book to within an inch of its life and their sanities. The community of the Bennington Bisexual Writers Union, Local

One—Benny Peterson and JP Solheim—proved invaluable. And speaking of Bennington, the Writing Seminars professors Sven Birkerts, Jenny Boully, Susan Cheever, and Peter Trachtenberg (and their workshops) cajoled, prompted, challenged, and encouraged this book. From Oberlin, Dorit Ganson taught me to think like a philosopher, and Laurie McMillan taught me to think like a writer; it may have been more than a decade ago (yikes!), but their classes shaped this book. In every book, the author goes through a dark night of the soul; with his feedback, Micah Lott pulled the sun up past the horizon during mine. Bill Powers let me believe I could write a book like this. Alice Ollstein let me borrow her eyes. Thank you, Sonia Roubini, for your moral clarity.

Despite their better judgment, my past and present colleagues at the *Wall Street Journal* have held my hand as I've learned how to report the news, fulfilling more than one childhood dream. I'm looking at Rob Barry, Chad Day, Anthony DeBarros, Theo Francis, Kirsten Grind, Coulter Jones, Susan Pulliam, Shalini Ramachandran, Sam Schechner, Mike Siconolfi, Georgia Wells, Brian Whitton, and too many others to name.

Family and friends listened to me complain at every step of this process, and I'm sorry, but I do love you. I'm sure I'm forgetting somebody, and if it's you, thank you.

-30-

SOURCES

Chapter 1: Internet

Benjamin, Walter. "Unpacking My Library." 1931.

Benton, Joshua. "Gone in a Flash: How Adobe's Abandonment of Flash for Mobile Devices Impacts News Orgs." *Nieman Lab* (blog), November 9, 2011. https://www.niemanlab.org/2011/11/gone-in-a-flash-how-adobes-abandonment-of-flash-for-mobile-devices-impacts-news-orgs/.

Carr, Nicholas. "Goodbye Rough Type, Hello New Cartographies." *Rough Type* (blog), October 10, 2024. https://www.roughtype.com/?p=9290.

Carr, Nicholas. "Who Killed the Blogosphere?" *Rough Type* (blog), November 7, 2008. https://www.roughtype.com/?p=1190.

Dash, Anil. "'Wherever You Get Your Podcasts' Is a Radical Statement." *Anil Dash* (blog), February 6, 2024. https://www.anildash.com//2024/02/06/wherever-you-get-podcasts/.

Del Mastro, Addison. "The Bones of the Blogosphere." *Bulwark* (blog), November 8, 2024. https://www.thebulwark.com/p/the-bones-of-the-blogosphere.

Duffy, Clare, and Kerry Flynn. "Some of the Most Iconic 9/11 News Coverage Is Lost. Blame Adobe Flash." CNN Business, September 10, 2021. https://www.cnn.com/2021/09/10/tech/digital-news-coverage-9-11.

Farrell, Maria, and Robin Berjon. "We Need to Rewild the Internet." *Noema*, April 16, 2024. https://www.noemamag.com/we-need-to-rewild-the-internet.

"Fire in Baltimore Snarls Internet Traffic, Too." Associated Press, July 20, 2001. https://www.nytimes.com/2001/07/20/us/fire-in-baltimore-snarls-internet-traffic-too.html.

Hass, Robert. *Time and Materials: Poems, 1997–2005*. Ecco, 2007.

Homer. *The Odyssey*. Translated by Emily Wilson. W. W. Norton, 2017.

Kalogeropoulos, Antonis, and Nic Newman. "'I Saw the News on Facebook': Brand Attribution When Accessing News from Distributed Environments." Reuters Institute for the Study of Journalism at the University of Oxford, 2017. https://doi.org/10.60625/RISJ-207C-K891.

Kelly, Jacques. "Shrouded in Darkness, Baltimore's Howard Street Tunnel Is a Workhorse Rail Artery." *Baltimore Sun*, July 27, 2019. https://www.baltimoresun.com/2019/07/27/shrouded-in-darkness-baltimores-howard-street-tunnel-is-a-workhorse-rail-artery/.

Lichterman, Joseph. "People Who Get News from Social or Search Usually Don't Remember the News Org That Published It, Survey Finds." *Nieman Lab* (blog), July 19, 2017. https://www.niemanlab.org/2017/07/people-who-get-news-from-social-or-search-usually-dont-remember-the-news-org-that-published-it-survey-finds/.

National Oceanic and Atmospheric Administration. "Climatological Data for Baltimore Area, MD." ThreadEx, July 2001.

National Transportation Safety Board. "Railroad Accident Brief." August 2001. https://www.ntsb.gov/investigations/AccidentReports/Reports/RAB0408.pdf.

Neely-Cohen, Maxwell. "Century-Scale Storage." Library Innovation Lab at Harvard Law School, 2024. https://lil.law.harvard.edu/century-scale-storage.

"Operating System Market Share Worldwide." StatCounter, July 17, 2025. https://gs.statcounter.com/os-market-share.

Pierce, David. "How Google Reader Died—and Why the Web Misses It More Than Ever." *Verge*, June 30, 2023. https://www.theverge.com/23778253/google-reader-death-2013-rss-social.

Schwartz, Delmore. "Calmly We Walk Through This April's Day." In *Selected Poems (1938–1958): Summer Knowledge*. New Directions, 1967. https://www.poetryfoundation.org/poems/42633/calmly-we-walk-through-this-aprils-day.

"Store." In *Online Etymology Dictionary*. Accessed August 13, 2024. https://www.etymonline.com/word/store.

Warren, Tom. "Huge Web Outage Takes Reddit, Twitch, and Other Big Sites Offline for an Hour." *Verge*, June 8, 2021. https://www.theverge.com/2021/6/8/22523953/twitch-reddit-down-fastly-outage-issues.

Chapter 2: Library

Albert, Kendra, and Jonathan Zittrain. "War and Nookd." *Future of the Internet* (blog), June 1, 2012. http://blogs.harvard.edu/futureoftheinternet/2012/06/01/war-and-nookd/.

Brodkin, Jon. "Nook Version of War and Peace Turns the Word 'Kindled' into 'Nookd.'" *Ars Technica*, June 1, 2012. https://arstechnica.com/information-technology/2012/06/nook-version-of-war-and-peace-turns-the-word-kindled-into-nookd/.

Chapekis, Athena, Samuel Bestvater, Emma Remy, and Gonzalo Rivero. "When Online Content Disappears." Pew Research Center Blog, May 17, 2024. https://www.pewresearch.org/data-labs/2024/05/17/when-online-content-disappears/.

@dox_gay. "Twitter hasn't noticed . . ." Twitter (now X), July 22, 2021. https://twitter.com/dox_gay/status/1418189401188970503.

Gault, Matthew, and Jason Koebler. "A Defunct Video Hosting Site Is Flooding Normal Websites with Hardcore Porn." *Motherboard*, July 22, 2021. https://www.vice.com/en/article/qj8xz3

/a-defunct-video-hosting-site-is-flooding-normal-websites-with-hardcore-porn.

Google. "How Google Search organizes information." Accessed December 3, 2021. https://www.google.com/search/howsearchworks/how-search-works/organizing-information/.

Google. "How Google Search works." Accessed May 22, 2024. https://www.google.com/search/howsearchworks/how-search-works/organizing-information/.

Google. "How Search organizes information." Accessed January 3, 2022. https://www.google.com/search/howsearchworks/crawling-indexing/.

Hern, Alex. "World's Most Delayed Software Released After 54 Years of Development." *Guardian*, June 6, 2014. https://www.theguardian.com/technology/2014/jun/06/vapourware-software-54-years-xanadu-ted-nelson-chapman.

Hitt, Tarpley. "BuzzFeed Quietly Deleted Thousands of Old Images." *Gawker*, October 26, 2021. https://www.gawker.com/media/what-happened-to-buzzfeeds-old-photos.

Hitt, Tarpley. "Pictures Disappear En Masse from G/O Media Posts." *Gawker*, October 27, 2021. https://www.gawker.com/media/go-media-staffers-confused-about-missing-photos.

"Human." Wikipedia, September 2, 2024. https://en.wikipedia.org/w/index.php?title=Human&oldid=1243615054.

"Internet Archive Headquarters." *Atlas Obscura*. Accessed July 17, 2025. https://www.atlasobscura.com/places/internet-archive-headquarters.

LaFrance, Adrienne. "The Internet's Dark Ages." *Atlantic*, October 14, 2015. https://www.theatlantic.com/technology/archive/2015/10/raiders-of-the-lost-web/409210/.

Lepore, Jill. "The Cobweb." *New Yorker*, January 19, 2015. https://www.newyorker.com/magazine/2015/01/26/cobweb.

Nelson, Ted. "The Xanadu Universe." Project Xanadu, 2016. https://xanadu.com/xUniverse-D6.

Nelson, Ted. "Xanalogical Structure." Project Xanadu, 2023. https://www.xanadu.com.au/ted/XUsurvey/xuDation.html.

Ortiz, Andi. "Disney+ Scrubs 'The Simpsons' Episode with Tiananmen Square Joke from Hong Kong Service." *TheWrap* (blog), November 29, 2021. https://www.thewrap.com/disney-scrubs-the-simpsons-episode-with-tiananmen-square-joke-from-hong-kong-service/.

Phillips, Heather. "The Great Library of Alexandria?" *Library Philosophy and Practice*, April 18, 2012. https://web.archive.org/web/20120418191647/http://unllib.unl.edu/LPP/phillips.htm.

Spayd, Liz. "Taking the Stealth Out of Editing." *New York Times*, September 24, 2016. https://www.nytimes.com/2016/09/25/public-editor/liz-spayd-new-york-times-public-editor.html.

Sullivan, Margaret. "Is The Times Being Stealthy? Or Just Improving Its Reporting in Real Time?" *New York Times*, October 14, 2013. https://publiceditor.blogs.nytimes.com/2013/10/14/is-the-times-being-stealthy-or-just-improving-its-reporting-in-real-time/.

Sullivan, Margaret. "Were Changes to Sanders Article 'Stealth Editing'?" *New York Times*, March 17, 2016. https://publiceditor.blogs.nytimes.com/2016/03/17/new-york-times-bernie-sanders-coverage-public-editor/.

"Wikipedia:Getting to Philosophy." Wikipedia, August 22, 2024. https://en.wikipedia.org/w/index.php?title=Wikipedia:Getting_to_Philosophy&oldid=1241627159.

Wolf, Gary. "The Curse of Xanadu." *Wired*, June 1, 1995. https://www.wired.com/1995/06/xanadu/.

Yglesias, Matthew. "Internet Triumphalism." *Think Progress* (blog), August 20, 2008. https://archive.thinkprogress.org/internet-triumphalism-c1204d6d022d/.

Zittrain, Jonathan. *The Future of the Internet—And How to Stop It*. Yale University Press, 2008.

Zittrain, Jonathan. "The Internet Is Rotting." *Atlantic*, June 30, 2021.

https://www.theatlantic.com/technology/archive/2021/06/the-internet-is-a-collective-hallucination/619320/.

Zittrain, Jonathan. "Orwellian Indeed." *Future of the Internet* (blog), July 17, 2009. http://blogs.harvard.edu/futureoftheinternet/2009/07/17/orwellian-indeed/.

Chapter 3: News

"The Awl Archive Is Gone." *Flaming Hydra*, July 26, 2024. https://flaminghydra.com/the-awl-archive-is-gone/.

Bolter, Jay. "A Vampiric Hedge Fund Is Eating the Pioneer Press Alive." *Racket* (blog), November 15, 2023. https://racketmn.com/alden-capital-vampiric-hedge-fund-st-paul-pioneer-press.

"City Pages." *City Pages*, October 29, 2020. https://www.citypages.com/.

Collins, Bob. "Big Profits at Pioneer Press as Corporate Parent Destroys It." NewsCut, May 1, 2018. https://newscut.mprnews.org/2018/05/big-profits-at-pioneer-press-as-corporate-parent-destroys-it/./index.html.

Darr, Joshua P., Matthew P. Hitt, and Johanna L. Dunaway. "Newspaper Closures Polarize Voting Behavior." *Journal of Communication* 68, no. 6 (2018): 1007–28. https://doi.org/10.1093/joc/jqy051.

Doctor, Ken. "Newsonomics: Alden Global Capital is making so much money wrecking local journalism it might not want to stop anytime soon." *Nieman Lab* (blog), May 1, 2018. https://www.niemanlab.org/2018/05/newsonomics-alden-global-capital-is-making-so-much-money-wrecking-local-journalism-it-might-not-want-to-stop-anytime-soon/.

Jennings, Jay, and Meghan Rubado. "Newspaper Decline and the Effect on Local Government Coverage." Annette Strauss Institute for Civic Life at the University of Texas at Austin, 2019.

Powers, William. *Hamlet's Blackberry*. HarperCollins, 2010.

Rodriguez, Richard. "Final Edition." *Harper's Magazine*, November 1, 2009. https://harpers.org/archive/2009/11/final-edition/.

Scocca, Tom. "New York City, July 17, 2012." *Awl*, July 17, 2012. https://www.theawl.com/2012/07/new-york-city-july-17-2012/.

Seward, Zach. "What Was Quartz?" *Zach Seward* (blog), April 7, 2025. https://www.zachseward.com/what-was-quartz/.

Shaker, Lee. "Dead Newspapers and Citizens' Civic Engagement." *Political Communication* 31, no. 1 (2014): 131–48. https://doi.org/10.1080/10584609.2012.762817.

Shechmeister, Matthew. "Ghost Pages: A Wired.Com Farewell to GeoCities." *Wired*, November 3, 2009. https://www.wired.com/2009/11/geocities/.

Sicha, Choire. "Quartz Tried Everything in 10 Years." *Intelligencer*, April 28, 2022. https://nymag.com/intelligencer/2022/04/quartz-media-history.html.

Tate, Ryan. "How to Blow $3.5 Billion." *Gawker*, October 26, 2009. https://www.gawker.com/5390282/how-to-blow-35-billion.

West, John. "Death by a Thousand Likes: How Facebook and Twitter Are Killing the Open Web." *Quartz*, November 9, 2015. https://qz.com/545048/death-by-a-thousand-likes-how-facebook-and-twitter-are-killing-the-open-web.

West, John. "Publishers Are Desperately Pivoting to Video—but They Should Be Standing Up to Facebook." *Quartz*, July 26, 2017. https://qz.com/1038396/publishers-are-desperately-pivoting-to-video-but-they-should-be-standing-up-to-facebook.

Chapter 4: Death

Beauchamp, Zack. "One of the most famous campus outrage incidents was totally misrepresented." *Vox*, November 5, 2019. https://web.archive.org/web/20210608183012/https://www.vox.com/identities/2019/11/5/20944138/oberlin-banh-mi-college-campus-diversity.

Beebe, Galen, Lauren Clark, Fontaine Capel, and John West. *King of That Also*. Et Cetera Gallery Press, 2015.

Beelen, Kaspar, Jon Lawrence, Daniel C. S. Wilson, and David Beavan. "Bias and Representativeness in Digitized Newspaper Collections: Introducing the Environmental Scan." *Digital Scholarship in the Humanities* 38, no. 1 (2023): 1–22. https://doi.org/10.1093/llc/fqac037.

Bruenig, Elizabeth. "Why Won't Twitter Forgive Suey Park?" *New Republic*, May 20, 2015. https://newrepublic.com/article/121861/suey-parkof-cancelcolbert-fame-has-stopped-fighting-twitter.

Dickinson, Emily. "[A word is dead]." In *Poems (Third Series)*. Edited by Mabel Loomis Todd. Boston, 1896.

Ding, Jennifer, Jan Diehm, and Michelle McGhee. "Can Data Die? Tracking the Lenna Image." *Pudding*, 2021. https://pudding.cool/2021/10/lenna.

Google Search Help. "How Search works with your activity." Accessed August 11, 2025. https://support.google.com/websearch/answer/10909618?hl=en.

Hass, Robert. "Meditation at Lagunitas." In *Praise*. HarperCollins, 1979. https://www.poetryfoundation.org/poems/47553/meditation-at-lagunitas.

Herrman, John. "Time Is a Privacy Setting." *Awl*, August 13, 2015. https://www.theawl.com/2015/08/time-is-a-privacy-setting/.

Matt, Frank, Joanna Stern, Rob Barry, John West, and Georgia Wells. "Investigation: How TikTok's Algorithm Figures Out Your Deepest Desires." *Wall Street Journal*, 2021. https://www.wsj.com/video/series/inside-tiktoks-highly-secretive-algorithm/investigation-how-tiktok-algorithm-figures-out-your-deepest-desires/6C0C2040-FF25-4827-8528-2BD6612E3796.

MIT Media Lab. "The Electome: Measuring Responsiveness in the 2016 Election." Accessed August 11, 2025. https://www.media.mit.edu/projects/the-electome-measuring-responsiveness-in-the-2016-election/overview/.

Ong, Walter. *Orality and Literacy: The Technologizing of the Word*. Routledge, 1982.

Patel, Vimal. "Colleges Are Losing Control of Their Story. The Banh-Mi Affair at Oberlin Shows How." *Chronicle of Higher Education*, October 31, 2019. https://www.chronicle.com/article/colleges-are-losing-control-of-their-story-the-banh-mi-affair-at-oberlin-shows-how/.

Putnam, Lara. "The Transnational and the Text-Searchable: Digitized Sources and the Shadows They Cast." *American Historical Review* 121, no. 2 (2016): 377–402. https://doi.org/10.1093/ahr/121.2.377.

Raymond, Eric S. "BCPL." The Jargon File. Accessed April 14, 2025. http://www.catb.org/jargon/html/B/BCPL.html.

Robinson, Marilynne. *The Death of Adam*. Picador, 1998.

Robinson, Marilynne. *The Givenness of Things*. Farrar, Straus and Giroux, 2015.

Robinson, Marilynne. "Language Is Smarter Than We Are." *New York Times*, January 11, 1987. https://www.nytimes.com/1987/01/11/books/about-books-language-is-smarter-than-we-are.html.

Romero, Oscar. "Solemnity of Our Lady of Mount Carmel." Sermon. Iglesia El Carmen, Santa Tecla, El Salvador, July 16, 1977. https://carmelitequotes.blog/2022/03/23/romero-16jul77/.

Serino, Kenichi. "Bringing Historical Black Newspapers into the Digital Age, Students Discover Their Past." PBS News, February 28, 2023. https://www.pbs.org/newshour/nation/bringing-historical-black-newspapers-into-the-digital-age-students-discover-their-past.

Thompson, Alex. "Journalists and Trump Voters Live in Separate Online Bubbles, MIT Analysis Shows." *VICE*, December 8, 2016. https://www.vice.com/en/article/journalists-and-trump-voters-live-in-separate-online-bubbles-mit-analysis-shows/.

Tran, Clover Linh. "CDS Appropriates Asian Dishes, Students Say."

Oberlin Review, November 6, 2015. https://oberlinreview.org/9055/news/cds-appropriates-asian-dishes-students-say/.

Wasik, Bill. "A.I. Is Poised to Rewrite History. Literally." *New York Times Magazine*, June 16, 2025. https://www.nytimes.com/2025/06/16/magazine/ai-history-historians-scholarship.html.

West, John. "The Birds, the Bees, and Multi-National Corporations." *12.2 Enviroanime*, 2001. https://kirney.tripod.com/ehome/ksee1.html.

Wikipedia. "Help:Talk pages." Last edited July 24, 2025. https://en.wikipedia.org/w/index.php?title=Help:Talk_pages&oldid=1302306203.

Yarasavage, Nathan. "African American Newspapers Added to Chronicling America." *Headlines and Heroes* (Library of Congress blog), February 28, 2022. https://blogs.loc.gov/headlinesandheroes/2022/02/african-american-newspapers-added.

Chapter 5: Attention

"Attention." In *Online Etymology Dictionary*. Accessed August 11, 2025. https://www.etymonline.com/word/attention.

Bennett, Joshua. "We." *Poetry*, May 2025. https://www.poetryfoundation.org/poetrymagazine/poems/1668136/we.

Caprioglio Panizza, Silvia. *The Ethics of Attention: Engaging the Real with Iris Murdoch and Simone Weil* (Routledge Studies in Ethics and Moral Theory). Routledge, 2022.

deBoer, Freddie. "The Temporal Copernican Principle." *Freddie deBoer* (Substack), September 6, 2024. https://freddiedeboer.substack.com/p/the-temporal-copernican-principle.

Foer, Franklin. "'Attention Is the Beginning of Devotion.'" *Atlantic*, May 9, 2019. https://www.theatlantic.com/technology/archive/2019/05/mary-olivers-poetry-captures-our-relationship-technology/589039/.

Graham, Ruth. "Meet the Nun Who Wants You to Remember You Will Die." *New York Times*, May 14, 2021. https://www.nytimes.com/2021/05/14/us/memento-mori-nun.html.

Greenwell, Garth. *Small Rain*. Farrar, Straus and Giroux, 2024.

Hayes, Christopher. *The Sirens' Call: Inner Life in the Age of Attention Capitalism*. Penguin, 2025.

Keats, John. "Ode on a Grecian Urn." 1819. https://www.poetryfoundation.org/poems/44477/ode-on-a-grecian-urn.

Kulwin, Noah. "The Man Who Invented the Pop-up Ad on the Original Sin of Advertising." *Intelligencer*, April 23, 2018. https://nymag.com/intelligencer/2018/04/ethan-zuckerman-inventor-of-pop-up-ad-interview.html.

Meyer, Angela. "Reading Fiction Is Addictive and Unhealthy, Says Corn Flakes Guy." *LiteraryMinded*, September 30, 2011. https://literaryminded.com.au/2011/09/30/reading-fiction-is-addictive-and-unhealthy-says-corn-flakes-guy/.

Oliver, Mary. "When Death Comes." In *New and Selected Poems*, vol 1. Beacon, 1992.

Oliver, Mary. "Wild Geese." In *Dream Work*. Penguin, 1994.

Oliver, Mary. "Yes! No!" In *White Pine*. Harcourt, 1994.

Petrow, Steven. "Mary Oliver's Poems Taught Me How to Live." *New York Times*, January 18, 2019. https://www.nytimes.com/2019/01/18/well/mary-olivers-poems-taught-me-how-to-live.html.

Plato. *Phaedrus*. Translated by Benjamin Jowett. Accessed August 11, 2025. https://www.gutenberg.org/files/1636/1636-h/1636-h.htm.

Postman, Neil. *Amusing Ourselves to Death: Public Discourse in the Age of Show Business*. Penguin, 1985.

Postman, Neil. *Technopoly: The Surrender of Culture to Technology*. Knopf, 1992.

Roberts, Molly. "Sam Bankman-Fried Doesn't Read. That Tells Us Everything." *Washington Post*, November 29, 2022. https://

www.washingtonpost.com/opinions/2022/11/29/sam-bankman-fried-reading-effective-altruism/.

Rorty, Richard. *Philosophy as Poetry* (Page-Barbour Lectures). University of Virginia Press, 2016.

Sacasas, L. M. "Re-Sourcing the Mind." *Convivial Society* (Substack), August 1, 2024. https://theconvivialsociety.substack.com/p/re-sourcing-the-mind.

Schumacher, Ernst Friedrich. *Small Is Beautiful: Economics as if People Mattered*. Blond and Briggs, 1973. Reprint, Harper Perennial, 1989.

Weil, Simone. *Gravity and Grace*. Translated by Arthur Wills. G. P. Putnam's Sons, 1952.

Chapter 6: Friction

Aquinas, Thomas. *Summa Theologica*. Translated by the Fathers of the English Dominican Province. Benziger Brothers, 1947. https://www.ccel.org/a/aquinas/summa/home.html.

Ashton, Pete. "Sitting in Stagram." 2015. https://art.peteashton.com/sitting-in-stagram/.

Bolter, Jay David, and Diane Gromala. *Windows and Mirrors: Interaction Design, Digital Art, and the Myth of Transparency*. MIT Press, 2005.

Curtis, Benjamin. "Google Isn't Just a Search Engine — It's a Literal Extension of Our Mind." *Nieman Lab*, September 14, 2018. https://www.niemanlab.org/2018/09/google-isnt-just-a-search-engine-its-a-literal-extension-of-our-mind/.

Environmental Protection Agency. "Indoor Water Use in the United States." 2008.

Google. *Google: 2024 Environmental Report*. 2024. https://www.gstatic.com/gumdrop/sustainability/google-2024-environmental-report.pdf.

Hass, Robert. "Meditation at Lagunitas." In *Praise*. HarperCollins, 1979. https://www.poetryfoundation.org/poems/47553/meditation-at-lagunitas.

Heaney, Seamus. "Blackberry-Picking." 1966. In *Opened Ground: Selected Poems, 1966–1996*. Farrar, Straus and Giroux, 1999. https://www.poetryfoundation.org/poems/50981/blackberry-picking.

Heffernan, Virginia. *Magic and Loss: The Internet as Art*. Simon and Schuster, 2016.

Lialina, Olia. "Once Again, The Doorknob. On Affordance, Forgiveness and Ambiguity in Human Computer and Human Robot Interaction." Keynote lecture. Rethinking Affordance Symposium, Akademie Schloss Solitude, June 8, 2018. https://contemporary-home-computing.org/affordance/.

Lialina, Olia. "Turing Complete User." *Contemporary Home Computing*, October 2012. https://contemporary-home-computing.org/turing-complete-user/.

McDonald, Henry. "Seamus Heaney's Last Words Were 'Noli timere', Son Tells Funeral." *Guardian*, September 2, 2013. https://www.theguardian.com/books/2013/sep/02/seamus-heaney-last-words-funeral.

Meyer, Robinson. "How So Many People Got Seamus Heaney's Last Words Wrong," *Atlantic*. September 4, 2013. https://www.theatlantic.com/technology/archive/2013/09/how-so-many-people-got-seamus-heaneys-last-words-wrong/279330/.

Microsoft. *2024 Environmental Sustainability Report*. 2024.

Monserrate, Steven Gonzalez. "The Staggering Ecological Impacts of Computation and the Cloud." *MIT Press Reader*, February 14, 2022. https://thereader.mitpress.mit.edu/the-staggering-ecological-impacts-of-computation-and-the-cloud/.

Norman, Donald A. *The Invisible Computer: Why Good Products Can Fail, the Personal Computer Is So Complex, and Information Appliances Are the Solution*. MIT Press, 1999.

Norman, Donald A. "Why Interfaces Don't Work." In *The Art of Human-Computer Interface Design*, ed. Brenda Laurel. Addison-Wesley, 1990.

Peters, John Durham. *The Marvelous Clouds: Toward a Philosophy of Elemental Media*. University of Chicago Press, 2015.

Plato. *Phaedo*. Translated by Benjamin Jowett. Accessed January 1, 2022. http://classics.mit.edu/Plato/phaedo.html.

Stevens, Laura. "Amazon Finds the Cause of Its AWS Outage: A Typo." *Wall Street Journal*, March 2, 2017. https://www.wsj.com/articles/amazon-finds-the-cause-of-its-aws-outage-a-typo-1488490506.

Turkle, Sherry. *The Second Self: Computers and the Human Spirit*. MIT Press, 2005.

Warren, Tom. "Microsoft Wants Three Mile Island to Fuel Its AI Power Needs." *Verge*, September 20, 2024. https://www.theverge.com/2024/9/20/24249770/microsoft-three-mile-island-nuclear-power-plant-deal-ai-data-centers.

Wayland-Smith, Ellen. *The Science of Last Things: Essays on Deep Time and the Boundaries of the Self*. Milkweed Editions, 2025.

West, John. "What Late Poet Seamus Heaney's Last Text Tells Us about Our Digital Lives." *Fast Company*, February 26, 2019. https://www.fastcompany.com/90311664/what-late-poet-seamus-heaneys-last-text-message-tells-us-about-our-digital-lives.

Winner, Langdon. "Technologies as Forms of Life." In *Ethics and Emerging Technologies*, edited by Ronald L. Sandler. Palgrave Macmillan, 2014. https://doi.org/10.1057/9781137349088_4.

Winner, Langdon. *The Whale and the Reactor: A Search for Limits in an Age of High Technology*. 2nd ed. University of Chicago Press, 2020.

Chapter 7: Algorithms

Ballhaus, Rebecca, Brody Mullins, Chad Day, John West, Coulter Jones, and James Grimaldi. "Federal Officials Trade Stock in Companies Their Agencies Oversee." *Wall Street Journal*, October 11, 2022. https://www.wsj.com/finance/regulation/government-officials-invest-in-companies-their-agencies-oversee-11665489653.

Day, Chad, John West, and Coulter Jones. "How WSJ Analyzed 12,000 Federal Officials' Financial Disclosures for Its 'Capital Assets' Investigation." *Wall Street Journal*, October 11, 2022. https://www.wsj.com/articles/how-wsj-analyzed-12-000-federal-officials-financial-disclosures-for-its-capital-assets-investigation-11665491120.

Foucault, Michel. *Discipline and Punish: The Birth of the Prison*. Vintage Books, 1975.

Giordano, Val. "How Many Books Are in the World?" *ISBNDB Blog*, October 20, 2023. https://isbndb.com/blog/how-many-books-are-in-the-world/.

Hobbs, Tawnell D., Rob Barry, and Yoree Koh. "'The Corpse Bride Diet': How TikTok Inundates Teens with Eating-Disorder Videos." *Wall Street Journal*, December 17, 2021. https://www.wsj.com/tech/how-tiktok-inundates-teens-with-eating-disorder-videos-11639754848.

Madrigal, Alexis C. "Google: There Are Exactly 129,864,880 Books in the World." *Atlantic*, August 5, 2010. https://www.theatlantic.com/technology/archive/2010/08/google-there-are-exactly-129-864-880-books-in-the-world/61024/.

O'Gieblyn, Meghan. "Routine Maintenance: Embracing Habit in an Automated World." *Harper's Magazine*, January 2022. https://harpers.org/archive/2022/01/routine-maintenance-embracing-habit-in-an-automated-world-meghan-ogieblyn/.

Ovid. *Metamorphoses: A New Verse Translation* (Penguin Classics). Translated by David Raeburn. Penguin, 2004.

Pelly, Liz. "The Ghosts in the Machine." *Harper's Magazine*, January 2025. https://harpers.org/archive/2025/01/the-ghosts-in-the-machine-liz-pelly-spotify-musicians/.

Pelly, Liz. *Mood Machine: The Rise of Spotify and the Cost of the Perfect List*. One Signal, 2025.

Postman, Neil. *Technopoly: The Surrender of Culture to Technology*. Knopf, 1992.

Price, Andy. "'More Music Is Being Released Today (in a Single Day) Than Was Released in the Calendar Year of 1989': How the Music Production Industry Has Taken Note of the Huge Number of Self-Releasing Artists." MusicRadar, November 14, 2024. https://www.musicradar.com/music-industry/more-music-is-being-released-today-in-a-single-day-than-was-released-in-the-calendar-year-of-1989-how-the-music-production-industry-has-taken-note-of-the-huge-number-of-self-releasing-artists.

Shakespeare, William. *The Tragedy of Macbeth*. 1623.

Spotify. "Spotify Debuts a New AI DJ, Right in Your Pocket." February 22, 2023. https://newsroom.spotify.com/2023-02-22/spotify-debuts-a-new-ai-dj-right-in-your-pocket/.

UCL (University College London). "The Panopticon." Bentham Project, May 17, 2018. https://www.ucl.ac.uk/bentham-project/about-jeremy-bentham/panopticon.

West, John. *Leaves of Grasstok*. 2025. https://leaves-of-grasstok.vercel.app/.

Chapter 8: Search

Brereton, Dmitri. "Google Search Is Dying." *DKB Blog*, February 13, 2023. https://dkb.blog/p/google-search-is-dying.

Christman, Phil. "'Skills' No. 'Activities,' Maybe." *The Tourist* (Substack), May 25, 2025. https://philipchristman.substack.com/p/the-scandal-of-the-particular.

Deng, Jia, Wei Dong, Richard Socher, Li-Jia Li, Kai Li, and Li Fei-Fei. *ImageNet: A Large-Scale Hierarchical Image Database*. n.d.

"Electronic 'Brain' Teaches Itself." *New York Times*, July 13, 1958. https://www.nytimes.com/1958/07/13/archives/electronic-brain-teaches-itself.html.

Gambino, Megan. "What Is on Voyager's Golden Record?" *Smithsonian Magazine*, April 22, 2012. https://www.smithsonianmag.com/science-nature/what-is-on-voyagers-golden-record-73063839/.

Gerstgrasser, Matthias, Rylan Schaeffer, Apratim Dey, Rafael Rafailov, Henry Sleight, John Hughes, et al. "Is Model Collapse Inevitable? Breaking the Curse of Recursion by Accumulating Real and Synthetic Data." arXiv, April 29, 2024. https://doi.org/10.48550/arXiv.2404.01413.

Hall, Nick (@inck). "Since this is a children's-ai-metaphor-correction account now: Using chatgpt to write a paper is not like using a car to get from point A to . . ." Bluesky, May 14, 2025. https://bsky.app/profile/inck.bsky.social/post/3lp5cvkqi6227.

Howe, Susan. *The Quarry*. New Directions, 2015.

Knausgaard, Karl Ove, Olivia Lasky, and Damion Searls. "The Reenchanted World: On Finding Mystery in the Digital Age." *Harper's Magazine*, June 2025. https://harpers.org/archive/2025/06/the-reenchanted-world-karl-ove-knausgaard-digital-age/.

Lefkowitz, Melanie. "Professor's Perceptron Paved the Way for AI—60 Years Too Soon." *Cornell Chronicle*, September 25, 2019. https://news.cornell.edu/stories/2019/09/professors-perceptron-paved-way-ai-60-years-too-soon.

Longpre, Shayne. "AI Crawler Wars Threaten to Make the Web More Closed for Everyone." *MIT Technology Review*, February 11, 2025. https://www.technologyreview.com/2025/02/11/1111518/ai-crawler-wars-closed-web/.

Mims, Christopher. "Huang's Law Is the New Moore's Law, and Explains Why Nvidia Wants Arm." *Wall Street Journal*, September 19, 2020. https://www.wsj.com/articles/huangs-law-is-the-new-moores-law-and-explains-why-nvidia-wants-arm-11600488001.

Mueller, Birgit, Chris Danis, and Giuseppe Lavagetto. "How Crawlers Impact the Operations of the Wikimedia Projects." *Diff*, April 1, 2025. https://diff.wikimedia.org/2025/04/01/how-crawlers-impact-the-operations-of-the-wikimedia-projects/.

OpenAI. "AI and efficiency." June 9, 2022. https://openai.com/index/ai-and-efficiency/.

Reisner, Alex. "The Unbelievable Scale of AI's Pirated-Books

Problem." *Atlantic*, March 20, 2025. https://www.theatlantic.com/technology/archive/2025/03/libgen-meta-openai/682093/.

Scocca, Tom. "The Dance of Ignorance." *Flaming Hydra*, January 11, 2025. https://flaminghydra.com/issue-235/.

Seetharaman, Deepa. "For Data-Guzzling AI Companies, the Internet Is Too Small." *Wall Street Journal*, April 1, 2024. https://www.wsj.com/tech/ai/ai-training-data-synthetic-openai-anthropic-9230f8d8.

Warzel, Charlie. "The Open Secret of Google Search." *Atlantic*, June 20, 2022. https://www.theatlantic.com/ideas/archive/2022/06/google-search-algorithm-internet/661325/.

Witt, Stephen. "How Jensen Huang's Nvidia Is Powering the A.I. Revolution." *New Yorker*, November 27, 2023. https://www.newyorker.com/magazine/2023/12/04/how-jensen-huangs-nvidia-is-powering-the-ai-revolution.

Chapter 9: Belief

Amnesty International. "Myanmar: Facebook's systems promoted violence against Rohingya; Meta owes reparations." September 29, 2022. https://www.amnesty.org/en/latest/news/2022/09/myanmar-facebooks-systems-promoted-violence-against-rohingya-meta-owes-reparations-new-report/.

Amnesty International. "Myanmar: The social atrocity: Meta and the right to remedy for the Rohingya." September 29, 2022. https://www.amnesty.org/en/documents/asa16/5933/2022/en/.

Centers for Disease Control and Prevention. "Coronavirus Disease 2019 (COVID-19): Frequently Asked Questions and Answers." March 8, 2020. https://web.archive.org/web/20200308000225/https:/www.cdc.gov/coronavirus/2019-ncov/faq.html.

DiFranco, Ani. "Self Evident." 2002. https://www.danah.org/ani/SoMuchShouting/SelfEvident.html.

Disha, Ilir, James C. Cavendish, and Ryan D. King. "Historical Events and Spaces of Hate: Hate Crimes against Arabs and Muslims in Post-9/11 America." *Social Problems* 58, no. 1 (2011): 21–46. https://doi.org/10.1525/sp.2011.58.1.21.

Douthat, Ross Gregory. *Believe: Why Everyone Should Be Religious.* Zondervan, 2025.

Dowie, Mark. "The Myth of a Wilderness Without Humans." *MIT Press Reader*, October 11, 2019. https://thereader.mitpress.mit.edu/the-myth-of-a-wilderness-without-humans/.

Gallagher, Mike. "Why Do Young Americans Support Hamas? Look at TikTok." *Free Press*, November 1, 2023. https://www.thefp.com/p/tik-tok-young-americans-hamas-mike-gallag.

Hofstadter, Richard. "The Paranoid Style in American Politics." *Harper's Magazine*, November 1964. https://harpers.org/archive/1964/11/the-paranoid-style-in-american-politics/.

"Improving How Meta Treats People and Communities around the World." Oversight Board, April 28, 2024. https://www.oversightboard.com/.

Klein, Ezra, and Ross Douthat. "Ross Douthat on Trump, Mysticism and Psychedelics." *New York Times*, April 25, 2025. https://www.nytimes.com/2025/04/25/opinion/ezra-klein-podcast-ross-douthat.html.

Klein, Naomi. *Doppelganger: A Trip into the Mirror World.* Farrar, Straus and Giroux, 2023.

Purnell, Newley, and Jeff Horwitz. "Facebook Services Are Used to Spread Religious Hatred in India, Internal Documents Show." *Wall Street Journal*, October 23, 2021. https://www.wsj.com/articles/facebook-services-are-used-to-spread-religious-hatred-in-india-internal-documents-show-11635016354.

"Removal of COVID-19 Misinformation." Oversight Board, April 20, 2023. https://www.oversightboard.com/decision/pao-sabu4p2s/.

Roose, Kevin, and Paul Mozur. "Zuckerberg Was Called Out Over Myanmar Violence. Here's His Apology." *New York*

Times, April 9, 2018. https://www.nytimes.com/2018/04/09/business/facebook-myanmar-zuckerberg.html.

Vosoughi, Soroush, Deb Roy, and Sinan Aral. "The Spread of True and False News Online." *Science* 359, no. 6380 (2018): 1146–51. https://doi.org/10.1126/science.aap9559.

Chapter 10: Walk

Herrman, John. "Time Is a Privacy Setting." *Awl*, August 13, 2015. https://www.theawl.com/2015/08/time-is-a-privacy-setting/.

Lepore, Jill. "The Cobweb." *New Yorker*, January 19, 2015. https://www.newyorker.com/magazine/2015/01/26/cobweb.

Williams, William Carlos. "This Is Just to Say." 1938. In *The Collected Poems: Volume I, 1909–1939*. New Directions, 1991. https://www.poetryfoundation.org/poems/56159/this-is-just-to-say.

Chapter 11: Read

Ashbery, John. "Untitled (Bridge Poem)." 1988. https://walkerart.org/magazine/john-ashbery-siah-armajani-bridge-poem.

Green, John. *The Fault in Our Stars*. Dutton, 2012.

Greenwell, Garth. "How Memorizing Poems Will Change Your Life." *To a Green Thought* (Substack), March 24, 2025. https://garthgreenwell.substack.com/p/memorizing-poems-will-change-your.

Greenwell, Garth. *Small Rain*. Farrar, Straus and Giroux, 2024.

Oppen, George. "Stranger's Child." *Poetry*, November 1960. https://www.poetryfoundation.org/poetrymagazine/browse?volume=97&issue=2&page=21.

Salmerón, Ladislao, Lidia Altamura, Pablo Delgado, Anastasia Karagiorgi, and Cristina Vargas. "Reading Comprehension on Handheld Devices versus on Paper: A Narrative Review and Meta-Analysis of the Medium Effect and Its Moderators." *Journal of Educational Psychology* (US) 116, no. 2 (2024): 153–72. https://doi.org/10.1037/edu0000830.

Schmelzer, Paul. “Not a Conduit but a Place: John Ashbery Reads His Poem for Siah Armajani’s Bridge.” Walker Art Center, September 12, 2017. https://walkerart.org/magazine/john-ashbery-siah-armajani-bridge-poem.

Schmelzer, Paul. “Under Wraps: What’s Happening with the Walker’s Siah Armajani Bridge?” Walker Art Center, May 18, 2018. https://walkerart.org/magazine/siah-armajani-irene-hixon-whitney-bridge-reconstruction/.

Chapter 12: Dream

Basile, Jonathan. Library of Babel (website). Accessed August 11, 2025. https://libraryofbabel.info/.

Beebe, Galen, and John West. *Kiss List*. Et Cetera Gallery, 2013. http://beebe-west.com/viz/kiss-list/.

Belcourt, Billy-Ray. *Coexistence: Stories*. W. W. Norton, 2024.

Borges, Jorge Luis. *The Library of Babel*. Etchings by Erik Desmazières. Translated by Andrew Hurley. David R. Godine, 2000.

Foer, Franklin. *World without Mind: The Existential Threat of Big Tech*. Penguin, 2018.

Hass, Robert. “Meditation at Lagunitas.” In *Praise*. HarperCollins, 1979. https://www.poetryfoundation.org/poems/47553/meditation-at-lagunitas.

Heaney, Seamus. “Blackberry-Picking.” 1966. In *Opened Ground: Selected Poems, 1966–1996*. Farrar, Straus and Giroux, 1999. https://www.poetryfoundation.org/poems/50981/blackberry-picking.

Klein, Ezra. “Beyond the ‘Matrix’ Theory of the Mind.” *New York Times*, May 28, 2023. https://www.nytimes.com/2023/05/28/opinion/artificial-intelligence-thinking-minds-concentration.html.

Melville, Herman. *Bartleby, the Scrivener*. 1853.

Merritt, Marcus J. (@marcusjmerritt.com). “A Borges story about a guy who gets AI to summarize all the world’s information for him, and then summarize the summary, until the AI has . . ." Bluesky, July 14, 2025. https://bsky.app/profile/marcusjmerritt.com/post/3ltwntr7btk2o.

@tripsnek. "Compressing the entire internet into interpolable vector space that can be queried in natural language remains actual sorcery that we skipped . . ." Bluesky, July 11, 2025. https://bsky.app/profile/tripsnek.com/post/3ltp5wpwoic2m.

West, John. *Lessons and Carols: A Meditation on Recovery*. Eerdmans, 2023.

Chapter 13: End

Calvin, John. *Institutes of the Christian Religion: The First English Version of the 1541 French Edition*. Translated by Elsie Anne McKee. Eerdmans, 2009.

Dellal, Pamela. "BWV 106 - Gottes Zeit ist die allerbeste Zeit (Actus Tragicus)" (Bach Translations). Emmanuel Music. Accessed July 18, 2025. https://www.emmanuelmusic.org/bach-translations/bwv-106.

Heller, Wendy. "'*Aus Eigener Erfahrung Redet*': Bach, Luther, and Mary's Voice in the *Magnificat*, BWV 243." *Understanding Bach* 10 (2015): 31–69.

West, John. *Homer*. 2025. https://internaut.club/.

INDEX

JOHN WEST is a writer and technologist. He reports the news with code at the *Wall Street Journal*, where his work has won multiple awards, including two Pulitzer prizes. He holds an MFA in writing from the Bennington Writing Seminars and degrees in philosophy and music from Oberlin College and Conservatory. The *Boston Globe* called his first book, *Lessons and Carols*, "lyrical and crystalline." He lives outside Kingston, New York, with his partner, children, and unruly cat.

Text set in Lyon Text, designed by Kai Bernau, with Irene Vlachou and Ilya Ruderman

Cover art: Erin Jia
Cover design: HD Brewer
Book design and composition: Lydia Christine Hall
Proofreader: Shannon Lee
Copyeditor: Victoria Emily Jones
Project editor: Laurel Draper
Acquiring editor: Lisa Ann Cockrel

Here Below champions ambitious books that eschew binaries between the sacred and profane. Learn more at herebelowbooks.com.